Typical RTLS Technologies

Typical technologies for indoor RTLS are

✔ **Locating at choke point:** Proprietary Passive (LF), Proprietary Active (LF), Rubee (LF), Proprietary Passive (HF), EPC Gen 2 Passive (UHF), EPC Gen 2 Semi-Passive (UHF), SAW (2.45 GHz), and Active RFID (UHF).

✔ **Presence-based locating:** Almost all technologies that can be used at choke points, room level, or locating precisely; although presence-based locating is a precise form of locating, you may want to implement it by using a minimal investment and/or by deploying as minimal a new infrastructure as possible.

✔ **Locating precisely:** Wi-Fi (2.4 GHz 802.11 b/g and 5.8 GHz 802.11a and 2.4–5.8 GHz 802.11n),

ZigBee (868 MHz, 915 MHz, 2.4 GHz), Ultrasound Active RFID (UHF), Proprietary (2.4 GHz), TV signal (50–800 MHz), UWB (3.1–10.6 GHz), and Bluetooth (2.4 GHz).

✔ **Locating by associating:** Almost all technologies that can be used for choke points, room level, or locating precisely.

✔ **Locating at room level:** Infrared (Optical 850 nm), Ultrasound (Ultrasonics), and Powerline (LF).

Typical technologies for outdoor RTLS are

✔ **Locating precisely:** GPS (1575.42 MHz/1227.60 MHz), cellular (UHF), WiMAX (2–66 GHz), and TV signal (50–800 MHz).

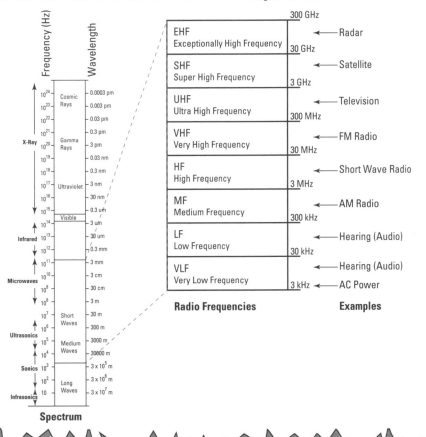

Location Models

- **Presence-based locating:** Location of a tag is returned whether or not it's present in a given area (such as a floor in a building).

- **Locating at room/sub-room level:** Location of a tag is returned as present in a specific room or as present in a specific area of the room.

- **Locating at choke points:** Location of a tag is returned as seen by a specific entry or exit point. By making use of strategically placed choke points and the time a tag was seen at a specific choke point, you can also determine the direction the tag is moving.

- **Locating by associating:** Location of a tag is returned as close proximity with respect to another tag.

- **Locating precisely:** Location of the tag is pinpointed precisely in the form of absolute coordinates, such as latitude, longitude, and altitude, or in the form of relative coordinates, such as distance in three dimensions from a reference point.

Top Security Threats

- **Physical destruction or theft of tags:** Because tags are wireless and small, they can easily be destroyed or stolen. Ensure that no data is stored in tags that can be retrieved easily by malicious hackers and that the tag has some form of tamper and heartbeat detection that tells you whether a tag is destroyed or removed from the asset.

- **Man in the middle:** In this attack, the attacker alters a legitimate message between the elements of RTLS (such as between the tag and the location sensor, the location sensor and the location engine, and so on) or deletes, adds to, changes, or reorders the message. Use airwaves noise-level baselines so that you can be alerted to any change in air traffic patterns. Also, use strong encryption, authentication, or technologies in which man in the middle attacks can't be done.

- **Denial of service:** The attacker may prevent or prohibit the normal use of air interface between the tag and the location sensor by disturbing the air interface, message flooding (sending repetitive messages), broadcast flooding, and so on to the location sensor or the tag. This can make the RTLS become very slow, become nonresponsive, or miss critical packets, causing the RTLS to not locate or to locate inaccurately. Use heartbeats to periodically check upon location sensors and to make use of some fixed tags that are always being located so that you know your RTLS system is operational and no active attack is in progress.

- **Eavesdropping:** The attacker monitors message exchanges between tags and location sensors or other elements of the RTLS. The eavesdropped information could, for example, be used to collect sensitive information about a person or to collect information from an asset. Use strong encryption, authentication, or technologies in which eavesdropping can't be done, and don't store sensitive information in the tags.

- **Crackers:** The attacker tries to *crack* (figure out) the security keys, if any, exchanged between the tags and the location sensors. By determining the security keys, the attacker gains access to the data network (or backbone network) of the facility. Use strong encryption, authentication, or technologies that can't be cracked.

For Dummies: Bestselling Book Series for Beginners

RTLS FOR DUMMIES®

by Ajay Malik

WILEY

Wiley Publishing, Inc.

RTLS For Dummies®

Published by
Wiley Publishing, Inc.
111 River Street
Hoboken, NJ 07030-5774

www.wiley.com

Copyright © 2009 by Wiley Publishing, Inc., Indianapolis, Indiana

Published by Wiley Publishing, Inc., Indianapolis, Indiana

Published simultaneously in Canada

For general information on our other products and services, please contact our Customer Care Department within the U.S. at 877-762-2974, outside the U.S. at 317-572-3993, or fax 317-572-4002.

For technical support, please visit www.wiley.com/techsupport.

Wiley also publishes its books in a variety of electronic formats. Some content that appears in print may not be available in electronic books.

Library of Congress Control Number: 2009924129

ISBN: 978-0-470-39868-5

Manufactured in the United States of America

10 9 8 7 6 5 4 3 2 1

WILEY

About the Author

Ajay Malik, a computer science graduate from IIT Roorkee (one of the most prestigious engineering universities in India), has over 20 patents issued and pending, many in the field of RTLS. He's been working in the field of RFID and RTLS for more than five years.

He works with a wide range of customers, vendors, and integrators for RTLS solutions in different market segments, such as health care, homeland security, education, industrial, and so on. Not only has he been involved in architecting and creating effective RTLS solutions by interacting with customers, but he also has been leading engineering teams to deliver components or complete RTLS solutions. He championed Real Time Location System, supporting multiple technologies at his tenure in Motorola and is currently working as CTO at RF Technologies, a company that has been involved with RTLS solutions for more than 21 years.

Dedication

To the three great women and a little man who define me every day — my mother Prabha, my wife Ritu, my daughter Shanaya, and my little son Aarush.

And, to my father, whom I could not get to tell all the things I had to say. He passed away too soon.

Acknowledgments

Let me take this opportunity to thank my family, friends, agent, and people at Wiley Publishing, Inc., whose support, dedication, and talent combined to make this book happen.

I want to thank my agent Matt Wagner, the best possible agent an aspiring book writer can have. He not only works hard to make things happen for you, but is also kind, encouraging, and very trustworthy. His sincere advice and efforts helped in getting this book underway.

At Wiley Publishing, I want to thank Katie Feltman and Jean Nelson for their encouragement and their ongoing support every step of the way. They made sure that the project stayed on course and made it into production so that all the talented folks on the Composition Services team could create this great final product. I also want to thank Colleen Totz Diamond, Kathy Simpson, and Jennifer Riggs for their manuscript reviews and updates and their invaluable input and suggestions on how best to present the content.

This book would not have been possible without the constant support from my wife, who constantly took over my responsibilities at home. She ensured that I had time available for this book. And my kids, who at the young ages of four and two somehow understood that "Daddy is busy." I also want to thank my brother Dr. Sanjay Malik, sister Alka Chawla, and friends Alan O'Brien, Mahender Vangati, Sarosh Vesuna, and Sameer Kanagala, who encouraged me all the time.

And, this acknowledgement section would be incomplete if I didn't mention Anthony Bartolo, Sujai Hajela, and Ray Martino, who introduced me to the world of RTLS; Glenn Jonas, CEO RF Technologies, Inc., who gave me the opportunity to try more in RTLS; and Terje "Terry" Aasen, Skip Speaks, Matt Perkins, Jarie Bolander, and Reik Read for providing me their views and inputs on different RTLS technologies. I sincerely thank them.

Publisher's Acknowledgments

We're proud of this book; please send us your comments through our online registration form located at http://dummies.custhelp.com. For other comments, please contact our Customer Care Department within the U.S. at 877-762-2974, outside the U.S. at 317-572-3993, or fax 317-572-4002.

Some of the people who helped bring this book to market include the following:

Acquisitions and Editorial

Project Editor: Jean Nelson

Senior Acquisitions Editor: Katie Feltman

Copy Editors: Jennifer Riggs, Kathy Simpson

Technical Editor: Sameer Kanagala

Editorial Manager: Kevin Kirschner

Media Development Project Manager: Laura Moss-Hollister

Media Development Assistant Project Manager: Jenny Swisher

Media Development Assistant Producers: Angela Denny, Josh Frank, Shawn Patrick, Kit Malone

Editorial Assistant: Amanda Foxworth

Sr. Editorial Assistant: Cherie Case

Cartoons: Rich Tennant (www.the5thwave.com)

Composition Services

Project Coordinator: Katie Key

Layout and Graphics: Shawn Frazier, Sarah Philippart, Christin Swinford, Christine Williams

Proofreader: Dwight Ramsey

Indexer: Potomac Indexing, LLC

Special Help
Colleen Totz Diamond

Publishing and Editorial for Technology Dummies

    Richard Swadley, Vice President and Executive Group Publisher

    Andy Cummings, Vice President and Publisher

    Mary Bednarek, Executive Acquisitions Director

    Mary C. Corder, Editorial Director

Publishing for Consumer Dummies

    Diane Graves Steele, Vice President and Publisher

Composition Services

    Gerry Fahey, Vice President of Production Services

    Debbie Stailey, Director of Composition Services

Contents at a Glance

Table of Contents

Introduction

*R*TLS *(Real Time Location System),* as the name implies, is all about location. You could buy this book because it was either present in a desired location or someone could locate it for you. Whether represented as a map, encoded as a ZIP code, labeled as a store aisle, or as any of the many other ways to represent location, people make fundamental decisions based on location. Location is a foundational information ingredient. If you need something, you need to know its location. If you want to show something, you need to put it in the right location. The ability to control or coordinate actions based on the location of things or people is fundamental. A business can set processes in place to ensure that specific assets are present in a specific location to increase sales. A professor can quickly account for a student's location at a time of distress. A business can ensure that workers can find shared equipment to prevent time wasted in searching for it. A paramedic can reach the trapped miner by taking the shortest route. A nurse can find the defibrillator in the shortest time. Businesses, schools, government, people, and so on can use location information in many ways to improve security, safety, service, Return on Investment (ROI), and in general, efficiency.

Today, with the increase in technological sophistication, it's now feasible to locate anything or anyone in real time anywhere. The system that accomplishes this is the *Real Time Location System (RTLS),* and this is usually achieved by making use of small electronic devices (or *tags*) attached to people or things at any time. You may have bought this book because there's something about an RTLS that you want to know more about, or maybe you bought this book only because it was in a *really* right location. Whatever the reason, I hope to give you the theories you need to know with the practical experience and advice you need to get an RTLS working for you.

About This Book

This book gives you the whole RTLS story, from soup to nuts, in the easy to understand *For Dummies* format. I begin with a tour of the RTLS applications and then serve you the details of what goes into an RTLS and what the various RTLS technologies are. Then, I plunge into the practical things that you need to consider before installing an RTLS and the common pitfalls that you need to be aware of. I include a chapter on evaluating an RTLS solution or selecting an RTLS vendor. And because RTLS security is equally as important as its accuracy, I also include a chapter on possible RTLS hacks and attacks.

After reading this book, you'll know what an RTLS is, how an RTLS is achieved, how to choose the right RTLS technology, and how you can make an RTLS work for you.

Like other *For Dummies* books, this one's been designed to let you skip around as much as possible. If you don't want to read this book cover to cover, you don't have to. Your current knowledge or experience of an RTLS makes some topics more relevant than others. If you want to zero-in on a particular topic, you can use the Table of Contents and index to focus on that coverage. As much as possible, I made it unnecessary for you to have to remember anything covered in another section of the book. From time to time, however, you'll come across a cross-reference to another section or chapter in this book. For the most part, such cross-references are meant to help you get more complete information on the subject, should you have the time and interest. If you have neither, no problem; just ignore the cross-reference as if it never existed.

Conventions Used in This Book

Some books have a dozen dizzying conventions that you need to know before you can even start. Not this one.

All you need to know is that new terms are given in *italics*.

When I discuss formulas (yes, I admit: I give a few formulas in the book), they're indented from regular paragraphs like so:

$$E = mc^2$$

Foolish Assumptions

Well, technically speaking, this book is *For Dummies*. However, you and I both know what that means. *For Dummies* is a style for reference books that introduces a technology, a concept, or a product in an easy to understand way.

This book is for anyone who wants to find out more about the RTLS and make use of real-time information to improve productivity, find and address deficiencies in processes, enhance customer support, or increase safety or security. Whether you're brand new to an RTLS or you're looking to expand

your current RTLS deployment, this book is for you. This book is for network administrators who are responsible for selecting, deploying, or maintaining the RTLS. This book is for the business solutions vendors as well as RTLS applications developers who are developing the RTLS applications and want to understand the landscape of RTLS technologies. And, this book is also for the innovators who want to create new technologies or improvise existing technologies to address what isn't addressed by current technologies.

For this book, I don't assume that you have in-depth knowledge of Wi-Fi, radio frequency identification (RFID), or any specific technology; however, this book does have some geek-speak from time to time. Like any *For Dummies* book, those sections are clearly marked with the Technical Stuff icon, which you can skip.

How This Book Is Organized

RTLS For Dummies is organized into six different parts. The following sections describe the various parts in this book.

Part I: Getting Your Bearings in RTLS

As the name implies, this part gets you started on your RTLS tour. Here, you get an overview of how an RTLS is used today and what it has to offer. You see a good sampling of the RTLS applications in Chapter 1 and get acquainted with the RTLS elements in Chapter 2. Chapter 2 also delves into various locating methodologies and technologies that can be used to achieve an RTLS. Then, you get familiar with all that's involved to implement the right RTLS in Chapter 3.

Part II: Implementing RTLS in Your World

This part is all about the practical advice that will help you make the RTLS work for you. You get guidance for planning your application and ideas for ways you might use the RTLS in your organization in Chapter 4. Chapters 5 and 6 deal primarily with preparing for installation and monitoring an RTLS. And, I explain the opportunities for integration of an RTLS with other business applications in Chapter 7.

Part III: Tag-A-Palooza: RTLS Technology Tour

In this part, you build the foundation that will help you choose the right locating technology. Whether you need to locate precisely, detect presence, or locate at any other granularity, Chapters 8–13 give you details on various technologies that can be used to make that happen.

Part IV: Monitoring Performance and Securing RTLS

In Part IV, I share with you the most important aspect of any system — performance monitoring and securing. Chapter 14 provides you with details and how to monitor performance of your RTLS as well as a few metrics that can be used to measure and present the performance of your RTLS system. Chapter 15 armors you with the knowledge of hacks and attacks on an RTLS so that you can establish appropriate defense mechanisms to protect the system's privacy and performance. Countermeasures are also discussed in Chapter 15.

Part V: The Part of Tens

No *For Dummies* book is complete without a Part of Tens. In this part, you find three chapters. In Chapter 16, I share with you my list of the ten most common pitfalls that you might run into — and what to do about them. Chapter 17 provides you ten tips for evaluating RTLS vendors. And because many tags require batteries, I provide ten tips on making the best use of batteries (proper charging, storage, and so on) in Chapter 18.

Part VI: Appendixes

When it comes to an RTLS, a single resource is never enough. Appendix A lists essential resources for staying up to speed on all things RTLS. Knowing where to find these resources will let you be effective in your RTLS choices. Appendix A also lists essential RTLS references.

Because when an RTLS is applied to people, it can be viewed as a potential threat to privacy, I've devoted Appendix B to privacy issues. And because it's critical to build a compelling business case to justify investments in an RTLS, I show you how to create fact-based benefits realization reports in Appendix C.

Icons Used in This Book

The following icons are strategically placed in the margins to point out stuff you may or may not want to read.

This Tip icon points out a handy shortcut or other valuable hints related to the topic at hand.

This icon marks something to remember — information that's especially important to know. To siphon off the most important information in each chapter, just skim through these icons.

This icon alerts you to nerdy discussions that you may well want to skip (or read when no one else is around).

Although the Warning icon appears rarely, when you need to be wary of a problem or common pitfall, this icon lets you know.

Where to Go from Here

All right, you're all set and ready to jump in. You can jump in anywhere you like — the book was written to allow you to just do that. But, if you want to get the full story from the beginning, jump into Chapter 1 first — that's where all the action starts. (If you're familiar with what RTLS is all about and are already in the process of evaluating various technologies, you might want to flip to Chapter 8.)

Part I
Getting Your Bearings in RTLS

The 5th Wave   By Rich Tennant

RICHTENNANT

PCS PHONES

"So does this phone have RTLS or GPS capabilities?"

In this part . . .

With all the buzz in recent years around RFID and then the success of GPS, the concept that you can locate anything or anybody automatically in real time is catching everyone's attention. In Chapter 1, I give you some examples of where an RTLS is, or could be, used. An RTLS can be the solution for applications in which items or personnel need to be located within a given area — whether they're moving or stationary. The examples I give get you thinking. The rest is up to you.

To get a good understanding of any RTLS, you have to know what it's made of. That's where Chapter 2 comes in, with the components, technologies, and methodologies of an RTLS. And because no single technology has been completely up to the task of providing an RTLS in all environments under all conditions, Chapter 3 lays down various things you might want to weigh before making your choice. From this humble beginning, it's a quick trip to total RTLS mastery.

Chapter 1

Getting to Know RTLS

A *Real Time Location System (RTLS)* enables you to find, track, manage, analyze, leverage, and otherwise use the information regarding where assets or people are located. Imagine being able to track people and assets at a local, national, or even global level. Retailers, military, law enforcement, emergency first response, healthcare, education, almost every business — even home life — can all benefit, or already do, by using RTLS technology. This chapter provides you with an overview of the RTLS — what it is and how it works — and describes who is using it and why.

Understanding How an RTLS Works

Location depicts or illustrates where something is well, located. With a *Real Time Location System,* or *RTLS,* you locate and track people and assets by associating a *tag,* a small wireless device, with each person or asset. *Assets* are items like laptops, projectors, equipment, and machinery. For example, you may attach a tag to office equipment (an asset) so that you can locate it within your office building; or you may carry an ID badge (a tag), and someone can locate you within the building.

Here are the parts of an RTLS, as shown in Figure 1-1:

✔ **Tags:** A mobile device that's enabled with location technology. A tag is usually small enough that it can be attached to assets or carried by people.

✔ **Location sensors:** Devices that usually have a known position. You use the location sensors to locate the tags that are affixed to the people and assets you want to track. For example, in the Global Positioning System (GPS), the satellites placed into orbit by the U.S. Department of Defense are the location sensors, and the location of tags (GPS receivers, such as the ones used in cars) is determined with signals from these satellites.

✔ **Location engine:** The software that communicates with tags and location sensors to determine the location of the tags. The location engine reports this information to middleware and applications.

✔ **Middleware:** The software that resides among the pure RTLS technology components (the tags, the location sensors, and the location engine) and the business applications capable of exploiting the value of the technology. You can think of middleware as the *plumbing.* Not only the specific RTLS application but also all other business applications used by the enterprise can be enriched with real-time location information.

✔ **Application:** The application (also known as *application software, end-user application,* or *software application*) is the software that interacts with the RTLS middleware and does work users are directly interested in. For example, an application that's always checking the location of kids in an entertainment park and then providing a Web page for parents to check the location of their kids.

The location engine, middleware, and application software may run on the same computer or on different ones. These applications also usually have a client interface such as a browser or PDA interface.

Chapter 2 covers the parts of an RTLS and how they work in much more detail. The following sections provide a more general overview of an RTLS to get you started.

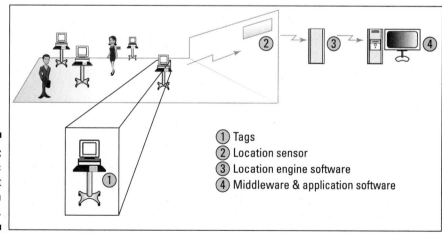

Figure 1-1:
The basic
parts that
comprise an
RTLS.

① Tags
② Location sensor
③ Location engine software
④ Middleware & application software

Using tags to locate people and assets

To locate people and assets in real time, an RTLS continually monitors the tag locations. Then the RTLS passes the location information to an application that makes use of the data. Here are just a few ways you can use RTLS tags to locate people and assets:

- ✔ Attach tags to your company laptops so that you can get alerts when the laptops leave the building without the authorized owner.
- ✔ Attach tags to your school projectors so that you can track which classes have the projector.
- ✔ Build a tag right into the machinery so that you can track its location in the factory.
- ✔ Require people to carry tags, such as

    - Badges
    - Wristbands
    - Pendants
    - Ankle bands
    - Belt clips

- ✔ Build tags into items that people already carry, such as cell phones.

Adding value with bells and whistles

Besides the technology that enables the location engine to locate the tags, tags often have bells and whistles that add tremendous value to the RTLS applications. These extra features are described in the following sections.

Push buttons

Tags with push buttons (or call buttons) can be used in multiple ways. One of the most common uses is as panic buttons that can be carried by people for summoning emergency response. Whenever the person presses the panic button, the location engine provides the alert and the location of the person pressing the button. Another use of push buttons in tags is when connected to assets, people can use the button to indicate status, such as Asset in Use, Work Order Completed, or Asset in Need of Repair.

Voice to voice

If a tag has voice-to-voice capability, you can use it to communicate with the person carrying the tag based on her location. For example, you can communicate location-based voice messages to a tourist in a museum, a visitor in an entertainment park, or a person trapped in any number of places.

Buzzers, LEDs, LCD screens, and vibrators

You can use buzzers, LEDs, or LCD screens on tags to communicate with the person carrying the tag, to identify or locate an asset, and to communicate with the person who has the asset or is expected to check the asset. Here's how you can use buzzers, LEDs, or LCD screens on tags:

- ✔ **Buzzers** can emit sounds, preprogrammed voice messages, ringtones, or live messages to give the tag carrier information or alerts.
- ✔ **LEDs** can provide the tag carrier with messages based on LED colors and blink patterns.
- ✔ **LCD screens** can display text messages for the tag carrier.

Sensors

You can incorporate many sensors in the tags to gain information about the environment, the status of the person carrying tag, or the asset that has the tag attached to it. Here are some examples of how sensors can help you work more effectively:

- ✔ **Motion sensors** in the tag can indicate whether the person carrying tag is moving.
- ✔ **A temperature sensor** attached to a painting can indicate whether that painting is in optimum thermal conditions.
- ✔ **A carbon monoxide sensor** can indicate whether the person carrying the tag is in a safe environment.

Connectors

Tags can also have connectors that connect to various assets to communicate specific details about the asset or its operation state. For example, the tag can indicate not only where the asset is but also whether the asset is powered on. You also can use the tag to turn the asset on or off.

Writeable memory

Tags can have writeable memory that can be used to store some user data for that asset, such as the name of the equipment renting company that placed the equipment.

Knowing the location models

When you want to know the physical location of an asset or a person, depending on your needs, you may want to know the absolute position, relative position, or symbolic position:

- **Absolute position** is the absolute coordinates, such as latitude, longitude, and altitude. For example, the Taj Mahal is situated at N27° 10' 0 N, E78° 2' 60, and the Washington Monument is at N 48.629283 and W –121.831533 with peak elevation of 4,455 feet (1,357.88 meters) above sea level.

- **Relative position** is the distance in three dimensions with reference to a fixed point. For example, the security guard is standing at 10 meters south of the main entrance of the building.

- **Symbolic position** implies presence in a specific area (for example, the doctor is in the operation theater room) or presence near something or someone (for example, the child is near her mom).

To satisfy the needs of various applications, whether they need precise location or room-level location, various RTLS systems report the location of tags in one of the following ways (see Figure 1-2):

- **Presence-based locating:** In this model, the tag location is returned as to whether it's present in a given area. For example, if your boss is carrying a tag, you can know whether he's in the building.

- **Locating at room level:** In this model, the tag location is returned as presence in a specific room. For example, if a schoolteacher presses the panic button to summon security assistance in the event of a classroom emergency, the location engine reports the teacher's exact classroom to the security guard.

- **Locating at sub-room level:** In this model, the tag location is returned as presence in a specific part of the room. For example, in hospital rooms that accommodate multiple patients, such as dual-bed rooms, if a nurse is carrying a tag, the location engine can report how much time the nurse spent by each patient's side.

- **Locating at choke points:** In this model, as shown in Figure 1-2, the tag location is returned by a specific *choke point* (an entry or exit point, such as a door; it's assumed that people or assets move from one area to another through these choke points). By monitoring the time a tag was seen at a specific choke point, you can also determine the direction the tag is moving. In other words, you can determine whether the tag is present inside or outside an area and whether it's entering or leaving the predetermined area. For example, if all visitors to a facility are required to carry tags, you can determine on which floor or building the visitors are located.

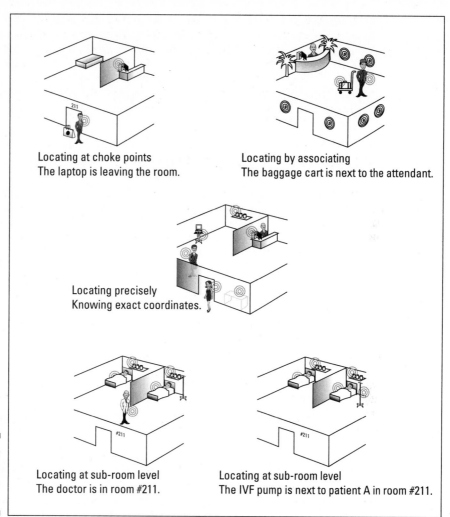

Locating at choke points
The laptop is leaving the room.

Locating by associating
The baggage cart is next to the attendant.

Locating precisely
Knowing exact coordinates.

Locating at sub-room level
The doctor is in room #211.

Locating at sub-room level
The IVF pump is next to patient A in room #211.

Figure 1-2:
RTLS
location
models.

✔ **Locating by associating:** In this model, tag location is returned as proximity with respect to another tag (see Figure 1-2). For example, if each patient in a clinic wears a tag and each IVF pump has a tag, the location of the IVF pump is returned as present next to a specific patient. The billing department can use this data, which indicates how long the IVF pump was in use with any patient, to calculate invoice amounts. Or maybe you need to know whether the owner of a specific laptop is actually next to that laptop. Many securities, financial, or other location-based services can benefit by knowing this type of information.

✔ **Locating precisely:** In this model, the tag location is pinpointed precisely. You can pinpoint the exact tag location on the map of the world and/or in a given building. The location is reported as absolute or relative position as described earlier. Locating precisely is the exact form of RTLS, and depending upon the accuracy of reporting, the precise location information can be extrapolated to room level, sub-room level, association level, presence level, and choke point.

Knowing the underlying technologies

Many systems and technologies have been developed over the years by researchers and commercial companies to provide the location of people, equipment, and other assets. Today, an RTLS can be achieved using light, camera vision, infrared, sound, ultrasound, Bluetooth, Wi-Fi, radio frequency identification (RFID), ZigBee, Ultra Wideband, GPS, Cellular, and many more technologies. These technologies are discussed in detail in Part III.

Different technologies use different approaches, and each approach solves a slightly different problem or supports different applications. These systems vary in many parameters, such as the physical phenomena used for location determination, the tag's form factor and location sensors, power requirements, range, indoor versus outdoor applicability, and time and space resolution. Some technologies determine location at room level, some can determine presence only, and some can determine location precisely. Some technologies work well outdoors whereas others are tailor-made for indoors. Some work well in an office environment, and others work well in an industrial environment. Some need additional location sensors, and some leverage existing infrastructures, such as electricity or Wi-Fi in the building.

In some systems, the tag being located actually computes its own position (also known as *tag self-positioning*); in some, the software that locates the tag is external to the tag (also known as *tag remote-positioning*); and in others, the tag position is determined by recognizing the location of a nearby tag (also known as *tag indirect-positioning*).

But all RTLS technologies share the common goal of computing the location of assets and people as accurately as needed by the application, and they all succeed in their own way.

Recognizing the Need for an RTLS

For the past few decades, the research, development, and standardization of radio frequency identification (RFID), Wi-Fi, and many other wireless technologies along with the success of GPS as a consumer navigation application have contributed to a dramatic increase in the number and type of users requiring the needs for location identification.

In addition to the developments in technologies that are making it increasingly feasible to identify and locate any thing or individual anywhere in real time, it's the pursuit of *what, how many,* and *where in real time* that's driving the need for RTLS. Whether you're talking about businesses, consumers, or public sector markets, the need for an RTLS is arising from the inherent need for just-in-time actionable information — the right information at the right time and location — which is a fundamental concept to make effective decisions and take immediate action:

✔ **Business markets:** The need for an RTLS is being driven by an ever more mobile workforce, the need for continuous information about worker and visitor safety, the economic pressures for operational efficiencies, aspirations to increase profit margins, the need to handle recalls effectively, demanding customers, and the desire for enhanced customer experience.

Customers are becoming less tolerant in accepting deficient products or services due to the wide range of choices available. For example, if a company can't track its shipments in real time, it likely can't deliver the kind of customer service its competitors offer. With an RTLS, customers can pick up the phone or check a Web site and find out instantly where their goods are at any given moment.

✔ **Consumer markets:** Concern with safety and security, need of location knowledge for social networking, aging population, and need for Web 3.0 are driving the need of an RTLS. People want to know immediately where children and seniors are located. Consumers and regulatory bodies want the ability to track the food from the moment it's picked in the field to the time it shows up at the grocery store so that they can pinpoint the source of accidental or malicious contamination of food and issue appropriate recalls. The youth and singles markets want to use location knowledge in the social network, and people expect a lot of location-based information from their personal navigation systems and when they surf the Web.

✔ **Public sector:** The concerns about terrorism and continued breakdowns in emergency communications during major disasters have opened the doors to RTLS applications as complements, alternatives, or backups to existing systems.

The applications that are driving the needs for an RTLS vary depending on whether you're talking about business, consumer, or public sector markets, the needs of individual segments within those markets, and the particulars of the targeted geographic region.

Tagging assets

Tagging is the act of affixing a tag to an asset or person. Tagging assets enables you to locate assets on demand, track and protect them, monitor their usage, trace them, and manage and plan inventory. Here are some examples of what you can accomplish by tagging assets:

- **Locate assets on demand:** On large sites, equipment easily goes missing. People forget to return it, and a lot of equipment gets moved and isn't returned to its original location. If the equipment is tagged, the RTLS can determine the equipment location. Such an application is especially useful in environments like hospitals where it's important to locate certain equipment quickly in the event of an emergency. Similarly, equipment requiring predictive or preventive maintenance or pieces of equipment for which a repair order or a recall has been issued can be located easily.

- **Track assets:** Sometimes you need to know not only the current location of an asset but also where it's been. This information is useful — or even critical — in many applications, for example, in hospitals when the staff needs to verify whether an asset has been through a disinfectant room after being in a patient room. Tracking can also be combined with sensors, such as a temperature sensor connected in the tag. In that case, you can't only determine where the sensor has been, but also see whether the asset has maintained or been stored at the right temperatures. For example, this can be used to capture information about the conditions the food is subjected to on the journey from the field to the grocer.

- **Protect assets:** Attaching RTLS tags to assets provides alerts when an asset moves closer to the perimeter of the facility or moves out of the facility. This kind of application is applicable for almost every business or government facility.

- **Monitor usage of assets:** Attaching RTLS tags to critical assets provides security officers the ability to track their movement and locate them at anytime. Tags also enable security to take corrective actions to ensure compliance with regulations when an unauthorized attempt to move that asset within a facility occurs, or even when the asset is taken from the facility. Furthermore, the storage of dangerous goods, such as explosives or weapons, can be made more secure. In this case, however, these goods can be additionally protected to ensure that only authorized personnel can be permitted to be within range of these assets. Alarms or other types of signals could be provided when the critical assets move without being close to the tags worn by the authorized personnel.

✔ **Trace assets:** An RTLS can help achieve the level of traceability, as needed by consumers, businesses, or policy makers. With an RTLS tag, you can record the location of a container or an item along with the time when it was seen at that location on the tag itself. This way, when a recall is issued for a specific item, it's easy to track the suspected path.

✔ **Improve asset utilization:** Analysis of equipment locations and length of time spent in particular locations can lead to an understanding of how often various pieces of equipment are used, where they're most often used, and what they're used for. The results of these analyses can help in deciding how much equipment is required, where to put the equipment to enable optimum work usage, and when to schedule equipment maintenance.

✔ **Manage and plan inventory:** One way to keep track of inventory is to subtract how many items you've used from how many you've ordered. The remaining difference is what you have in inventory. Whether because of human error, delay in entering information in inventory management systems, scrap, or stolen goods, data from books for inventory counting is rarely 100-percent accurate. Many organizations prefer to use a physical count of their materials. This is a great way to manage inventory if you're a handmade violin manufacturer and you make one violin a week. If, however, you're an equipment rental company dealing with tens of thousands of pieces of equipment and have constant turnaround, a physical count might not be so simple. Or, imagine if you had over 100 different types of products and maintaining optimal inventory levels for different types of products was an essential part of your business. Attaching tags to assets can help you in inventory reorder calculations, warehouse management, and some degree of inbound material planning.

Tagging people

You can benefit from tagging people in many ways. By tagging people, you can locate them on demand, protect and guide them, monitor their movement or activities, and provide emergency response. Here are some real-world examples of what you can accomplish by tagging people:

✔ **Locate people on demand:** In hospitals, tagging babies and children enables staff to locate them within the hospital and generate appropriate alerts should the child be abducted or otherwise go missing.

✔ **Protect and track people:** One of the highest risks within a senior care or assisted-living environment is a resident escape. A resident escape can cause harm to the resident and can be an enormous liability for the facility. An RTLS can help in resident tracking, and when a protected resident attempts to exit a facility, doors can be locked automatically and appropriate alerts can be triggered.

✔ **Monitor people's movements:** Many unique challenges can be solved by monitoring people's movements. For example, by monitoring security guards, an RTLS can help ensure that facility surveillance isn't hampered by a security guard sleeping on duty. Another example is in a mall or a city where security forces can be dispatched on detection of an unusual event, such as large number of people in smaller areas.

✔ **Provide emergency response:** By making use of push-button tags, the tag wearer can summon assistance when faced with an emergency situation. This can be used by teachers or students in schools, nurses or patients in hospitals, lone workers working in hazardous environments, or police officers requesting other police officers in man-down events.

✔ **Manage evacuations:** In events requiring evacuation, such as a fire, emergency first response can determine who's evacuated and who hasn't been evacuated.

✔ **Police restricted areas:** By attaching the tags to criminals, employees, or visitors, areas that are restricted can be better policed and access can be allowed or disallowed based on privilege levels. Furthermore, at any given time, security officers can monitor the whereabouts of everybody and take appropriate actions when necessary.

✔ **Improve workflow:** By analyzing trends of the utilization of assets and how workers and visitors spend their time, significant improvements in work practices and methodologies can be achieved. For example, in an emergency department, you can determine the key changes in equipment, staff, or processes required to decrease the time from the decision to admit a patient to the physical placement of the patient in an inpatient bed.

✔ **Improve customer service and response times:** By analyzing trends of time the visitors are forced to spend at specific areas, appropriate customer service agents can be added at appropriate locations in the facilities. By rearranging the customer service agent count, the facilities can ensure that optimum levels of customer service are provided throughout the facility.

✔ **Improve structure of facilities:** By analyzing trends of visitor dwell times, the facilities can be better structured for maximizing the facility's purpose. For example, in a painting exhibition, if some parts receive dwell times significantly more than others, the temporary structures can be rearranged to improve the flow of traffic through the exhibition.

Putting an RTLS to Work

Because RTLS technology has the potential to provide radical improvement and advancements to efficiency beyond the ordinary scope of improvement, many sectors are currently using an RTLS. The following sections describe just a handful of these sectors and how they're using an RTLS.

Healthcare

Hospitals, clinics, nursing homes, and other healthcare providers are looking at an RTLS to improve the bottom line and the care delivery potential. An RTLS can help

- **Locate healthcare personnel.** Quickly locating healthcare personnel is critical when a patient or staff member summons assistance during an emergency medical situation.

- **Track the movements of patients.** Tracking the physical movement of patients helps ensure patient safety. This is especially important for the safety and security of Alzheimer's and dementia patients. An RTLS can alert staff and give the location of a resident who wanders too far from a designated area or tries to leave the premises (known as an *elopement*) as well as when a patient passes too close to an entrance or an exit. Another example is infant abduction prevention in which an RTLS enables immediate infant location tracking, mother/baby matching, automatic door locking, and so on.

- **Improve throughput management.** Tracking patient flows for throughput management can potentially address problems, such as extended waiting times, overcrowding and boarding in the emergency department (ED) and post-anesthesia care units (PACUs), bumped and late surgeries, and lack of available routine and intensive care unit (ICU) beds. For example, using an RTLS reduces waiting and transfer time for patients because it may require less time to find staff or a wheelchair to transport the patient. Using an RTLS also enables close synchronization of housekeeping with patient discharge, enabling a faster bed turnaround rate.

- **Track equipment.** Tracking expensive or shared equipment, such as infusion pumps, saves time and helps staff to more easily find equipment that's recalled or due for preventive maintenance. Tracking equipment more quickly also improves inventory control, which reduces rental and purchasing costs.

✔ **Improve productivity of nurses and caregivers.** Because an RTLS can automate many tasks on the basis of location and in real time, it can reduce many mundane and repetitive tasks that nurses and caregivers encounter on a daily basis. For example, a nurse or a caregiver typically has to go into a room to cancel a call or trip a registry light manually — which often involves in tripping over extra chairs, patient visitors, and other equipment — but an RTLS can perform the same task automatically, simply by recognizing the nurse's presence in the room.

✔ **Improve patient/family satisfaction.** An RTLS can improve a patient's family satisfaction by increasing their awareness of patient location.

✔ **Improve staff safety.** Nurses face acts of violence, threats to safety, exposure to abusive language, and aggressive behavior from staff, patients, and visitors. An RTLS can become an important part of the solution by giving nurses a tool to request emergency assistance during a crisis.

✔ **Track personnel.** Tracking personnel alleviates security concerns by monitoring unauthorized access in restricted areas.

An RTLS is important for healthcare management personnel, other healthcare personnel, managed services providers, and those who are looking to solve complex issues in their respective healthcare organization.

Manufacturing

Manufacturers are always trying to do more with less. Manufacturers use RTLS technology to

✔ **Find products.** Often valuable production time is lost because the pallet with the right materials isn't where it should be and a search must be carried out. Armed with an RTLS and status information, manufacturers can reduce search time, which improves productivity and material flow while ensuring higher asset utilization.

✔ **Track progress.** At any given time, a manufacturing plant can contain thousands of products in various states of assembly. An RTLS can instantly identify what still needs to be done to each product. Historically, an alternative to having an RTLS was to have workers use bar codes or paper to track the stage of the product. One major problem with such an approach is that many employees forget to scan or enter the data.

✔ **Trace parts.** Using an RTLS enables manufacturers to trace parts, which can be used for recalls or investigations.

✔ **Track production items.** Many manufacturing processes are lengthy and have hundreds of steps that take place often in several different factory buildings. Being able to track production items accurately across a plant and link the location to the manufacturing stage results in fewer errors and improved workflow.

✔ **Find tools.** Manufacturers often need to tag shared tools so that they find them when needed without slowing production.

✔ **Locate works in process.** A *work order* is an order received by an organization from a customer or a client or is created internally within the organization. In a manufacturing environment, a work order is converted from a sales order to show that work will begin on the manufacturing, building, or engineering of the products requested by the customer. Tagging work orders can enable you to locate at what stage of manufacturing that specific work order is. A supervisor can also use some LEDs or other indicators on tags to communicate urgency levels for specific work orders.

The interaction between workers, machines, tools, work areas, and the products they're manufacturing has a significant location component. Whether in automotive, aerospace, computers, semi-conductor, or discrete manufacturing, RTLS technology can deliver improved quality at a reduced cost to make the manufacturing operations more competitive. An RTLS provides context and location-aware support to workers improving quality by reducing errors (and therefore reducing the cost of fixing them) and increasing efficiency by reducing process execution times. The devices and tools used by the workers can be made to know where they are and in which relation they stand to the production materials.

Automotive

The automotive industry has been one of the early adopters of RTLS technology. An RTLS helps in material logistics, vehicle management, and during manufacturing. Here are just a few tasks the auto industry accomplishes by using RTLS technology:

✔ **Reduce the cost of quality:** Auto makers reduce the cost of quality by preventing a vehicle with quality problems from being shipped while still shipping other vehicles. This can be achieved by using an RTLS to track information of finished vehicles from pre-delivery to shipment.

✔ **Manage supplies:** An RTLS locates critical tools and test equipment, and uses call buttons for replenishment of supplies.

✔ **Manage vehicles:** An RTLS reduces the average dwell times for finding vehicles by improving in-line/off-line vehicle tracking, vehicle (yard) management, dealer lots, and resale auction lots.

Aerospace and defense

Significant opportunities exist for an RTLS to help aerospace and defense. By using an RTLS, you can

✔ **Manage supplies.** An RTLS locates critical tools and test equipment, and uses call buttons for replenishment of supplies.

✔ **Fleet maintenance management.** An RTLS can be used to manage check-in and check-out processes.

✔ **Track maintenance dates.** An RTLS is flexible enough to track maintenance dates for critical support equipment and trailers across thousands of feet of open flight line or even while they move through metal cages in a warehouse.

✔ **Speed throughput.** An RTLS can perform spot locating and auditing capabilities to speed up throughput at critical transit points.

Research and development

Safeguarding research prototypes or products in early stage of development has been always a concern for research and development (R&D) facilities. Here are some tasks you can accomplish with RTLS technology in R&D:

✔ **Monitor and protect prototypes:** RTLS can enable the R&D teams to effectively monitor, inventory, or locate their prototypes, and have processes in place when an unauthorized attempt to move a prototype within a facility occurs. Prototypes can also be protected to ensure that only authorized personnel can be permitted to be within range of these prototypes.

✔ **Enforce escorting visitors:** Use an RTLS to enforce how visitors will be escorted. For example, you can specify who will escort visitors or how far away from an employee a visitor can be before an alert is triggered.

✔ **Police restricted areas:** By using RTLS tags, R&D facilities can distinguish between personnel and visitors and then track and locate them within a facility. Through the use of an RTLS, areas that are restricted can be better policed.

Retail

Marketing and customer service are prime concerns for the retail industry. RTLS technology helps improve performance in these areas by enabling retailers to do the following:

✔ **Understand consumer behavior:** By attaching tags to carts, retailers can obtain valuable data about consumer behavior and use this information to rearrange retail merchandise.

✔ **Improve customer service:** By having store employees carry tags, retailers can have the nearest located customer service agent assist customers when they request assistance.

✔ **Improve sales:** By attaching tags to products, retailers can have automated shelf management processes set up where the clerks in a store room can be notified immediately when the inventory of a specific item on the shelf goes below a specific level.

Mining

By using RTLS technology, people in tunnel and mining operations can improve safety. For example, they can

✔ **Track worker location.** With an RTLS, tunnel and mining operations can track the location of workers and record them in real time so that in the event of a disaster, everyone's most recent locations can be known immediately. You get a snapshot of where everybody was immediately before the disaster.

✔ **Enforce worker safety.** With an RTLS, you can ensure a full and functional set of protective and safety equipment is carried by all workers before entering a mine as well as all the time.

✔ **Assist emergency first responders.** The RTLS helps first responders to effectively locate the trapped miners and determine routes to the nearest safe exits.

An RTLS can also provide financially rewarding asset-tracking possibilities where the location of tools and vehicles can be available in real time to achieve improved asset utilization.

Visitor information

Visitor identification tags have long been used to provide security for office and industrial facilities. Because these tags can't be located from a distance, the effectiveness of these tags is limited by the need to escort the visitor at all times, which may not be practical. With an RTLS, you can

- ✔ **Track and locate personnel and visitors.** By using RTLS tags, security officers can identify who's within each section of a facility and track their movements to ensure that compliance with security regulations is maintained at all times.

- ✔ **Police restricted areas.** If visitors and employees wear RTLS tags, areas that are restricted can be policed better.

- ✔ **Ensure safe evacuations.** RTLS technology can also help address another challenge for the facilities — the safety and security of all visitors during unforeseen evacuations. Rescue personnel usually have well-defined systems to account for employees, but the records for people visiting the facility are usually too scattered or available only in lobby areas that may not be accessible during fires or other emergencies. If visitors wear tags, they enable a more targeted search and rescue operation in case there are visitors who might not have made it out of the building.

Industry conferences, amusement parks, and temporary setups

Industry trade shows, amusement parks, and county fairs have thousands of pieces of equipment that are set up temporarily, thousands of attendees, security staff, cleaning personnel, administrators, and other support staff. With an RTLS to help manage busy events, companies can

- ✔ **Track and communicate with people attending the event.** An RTLS enables the companies or other attendees to locate and contact attendees, exhibitors, administrators, or other support people while they roam throughout the conference facility.

- ✔ **Monitor equipment status.** With an RTLS, companies can more easily take inventory and monitor the location of all temporary equipment.

✔ **Improve booth layouts and set exhibit space purchase price.** By analyzing the trends of visitor dwell times in past shows, the exhibition organizations can modify the booth layouts and price exhibit space based on the dwell times it can potentially receive.

✔ **Locate missing persons.** Security staff can use an RTLS to help people locate their lost group member, such as a child.

Public safety

One of the primary government roles is providing safety and security for its citizens, property, critical infrastructure, and natural resources. Natural disasters, crimes, emergencies, terrorist attacks, and the threat of terrorist attacks require rapid response from public safety agencies, such as firefighters, police, and so on. An RTLS enables many applications that help people do their jobs better and save their own lives:

✔ **Determining man down:** Should a responder be struck down with an injury or unconsciousness (a *man-down event*), the RTLS can alert the main emergency response personnel with an accurate location of that person. The alarm can be activated by making use of one or more sensors, such as lack of motion, shock, and so on.

✔ **Locating firefighters:** The RTLS can guide searchers to locate firefighters in trouble.

Education

Terrorism and violence in schools and universities have added notable priority on emergency and first response services in schools. The increased popularity of online social networking has created the demand for real-time location, as well. With an RTLS, you can

✔ **Assist first response teams.** By using an RTLS, first response teams can more easily locate the teacher or student requesting assistance. With an RTLS, you also enable the teachers or security to target communication with the students or staff located in the distress area.

✔ **Identify people's locations.** By using an RTLS, participants in a social network on campus can find each other's location. This way, you can receive alerts when a friend is within a certain distance or get advice from people in the same locale.

Chapter 2

Knowing the Many Parts
of an RTLS

*W*hen most people consider an RTLS, they immediately think of the RTLS tags, which are the most tangible part of an RTLS that most people deal with. An RTLS includes a great deal more than tags, however. An RTLS also has location sensors, locating technology and methodology, location engine software, the middleware, and the application. Each of these components must be versatile to fit the requirements of companies with different environments, different needs, and different applications. This chapter shows how each part of an RTLS works and offers some ideas for how you can put an RTLS to work for your business.

Starting with the All-Important Tag

An RTLS *tag* is a miniature device that's enabled with location technology. You can use tags to locate the assets and the people who carry them. Tags must be small enough to be attached to assets or to be carried by people.

When you attach a tag to assets, such as office equipment, you typically glue the tag to the asset. Sometimes, people choose to screw or build the tag into assets, kind of like laptop computers that have built-in Wi-Fi, as opposed to a wireless network card. When you need to track people, you can have them carry tags as badges, wristbands, pendants, ankle bands, or belt clips.

The location of assets or people is nothing but the location of these tags. If the tag becomes separated from the asset or person, you can't track the asset or person.

Characterizing tags

Based on the tag's communication method with the other parts of the RTLS (including the location sensors and the location engine, which are covered later in this chapter), tags are characterized as passive, semipassive, or active. The following sections cover each type of tag, and Table 2-1 describes the specifications.

Passive tags

Passive tags are typically passive *radio frequency identification (RFID)* tags. A typical passive RFID system includes tags (or *transponders*) and interrogators (or *transceivers* or *readers*). The interrogator sends a radio signal that's received by the passive tags present in the RF (radio frequency) field of the interrogator. Tags receive the signal via their antennas and then respond by transmitting their stored data, as shown in Figure 2-1. Passive RFID tags have no battery and obtain the operational power to transmit data from the RF field emitted by a corresponding interrogator. This means that they can be read only when they are located in the RF field of the interrogator. The size and shape of the RF field depends upon the interrogator's antenna and starting transmit power. Typically, the field is about 10 meters. An interrogator can be an unattended standalone unit, such as for monitoring a door, or integrated with a mobile device, such as a handheld PDA.

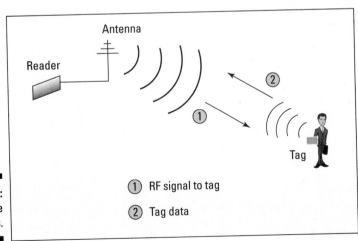

Figure 2-1:
Passive
RFID tags.

The tag circuit is designed in such a way that the tag operational power requirements are as low as 200 microwatts so that it can work on the power that can be absorbed from the RF field emitted by the interrogator. When a tag enters the RF field generated by the interrogator and receives sufficient energy, it becomes operational. In the operational state, the tag doesn't send a traditional RF transmission, but instead communicates back to the interrogator with a technique known as *backscatter*. In the backscatter technique, the tag reflects the carrier wave and changes the load on its antenna to reflect back an altered signal with information corresponding to the data stored on the tag such as tag ID. The interrogator's job then is to detect this change in the reflected signal. For these reasons, the passive tags usually have a short range, are light, have small form factors, and are inexpensive.

Surveillance in clothing stores and libraries, animal tracking, electronic toll collecting, and secure access control are a few examples of applications that use passive tags.

Semipassive tags

Semipassive tags (or *battery assisted passive tags*) are very similar to passive tags. They don't initiate any communication, need to be in the RF field of the interrogator to be read, and send data to the interrogator using the same backscatter technique as passive tags. However, semipassive tags have a small battery. The battery's main purpose is to either monitor environmental conditions or to offer greater range and reliability than passive tags. Note that the battery on semipassive tags isn't used to generate RF energy.

Because these tags contain more hardware than passive tags, they are typically more expensive. The semipassive tags can be used wherever passive tags are used; however, they're generally reserved for costly items or wherever sensors are also needed.

Active tags

Active tags, unlike passive or semipassive tags, contain an onboard radio (transmitter or transceiver) and are typically powered by an internal battery. Because they have onboard radio, they usually have a long range and can communicate without being prompted. For these reasons, these tags can be located in real time, say every second; however, the battery life duration depends strictly on the application. For example, if you want to attach tags to portable projectors to know their location, you want those tags to transmit only *once* whenever the projector stops moving. This way, you always know the projector's last position. However, if you also wanted to use the tags to not only know the location of projectors but also to get an alert if the projector leaves the building, you do want this tag to transmit *constantly* at high rates, such as twice a second when moving, so you can know the location in real time and take appropriate action immediately. Because many more transmits are involved, the tags consume more battery power in the latter application than the former.

Such tags are usually larger and more expensive than passive or semipassive tags. Because these tags have batteries, battery life is an important concern. Tag batteries are covered later in this chapter, in the section, "Recognizing the importance of a tag battery."

Specifications for tags

Table 2-1 describes the specifications for passive, semipassive, and active tags.

Table 2-1	Passive, Semipassive, and Active Tags		
Specifications	*Passive*	*Semipassive*	*Active*
Size	Depends on the desired range; from a grain of rice (range of less than 10 ft.) to a car license plate (range up to 50 ft.)	Depends on battery life and size; comparable to passive tags in size	Depends on battery life and size (for example, coin cells, AA, or AAA batteries)
Locatability	Only when interrogated (in the RF field of interrogator)	Only when interrogated (in the RF field of interrogator)	Locatable at a preset interval, such as every second or 10 seconds
Battery life	N/A	Up to 10 years	Up to 5 years
Range	Depends on the function of reader power and tag antenna size; usually 20 feet but can be up to 600 feet	Same as passive tags; up to 100 meters	Typically 100 meters but can be up to several hundred meters
Cost	A few cents	A few dollars	Tens of dollars
Processing capability	Almost none	Ability to read sensors, such as temperature	Much like very small computers

Considering tag ergonomics

The way that a tag is attached to assets or carried by people is *tag ergonomics.* When a tag is carried by people, good tag ergonomics means that a tag is specifically designed to be comfortable and easy to use, physically and psychologically. Tags that are attached to assets must be easy to attach. When attached to the asset, the tag must not interfere with the use or storage of the asset. The tag also must be easy to replace.

Here are two key factors to consider with tag ergonomics:

- **Size and shape:** You may not be able to use tags that are the same size and shape when you use them for multiple uses. For example, you may need a tag that can be attached to thermometers as well as attached to large hospital beds. In this case, you need to have tags of different sizes and shapes that work with the same RTLS.

- **Mounting options:** How a tag is mounted on an asset or the different ways that people can carry a particular tag type makes a difference in how you can use the tags. For example, a tag that needs a one-inch square surface to attach to assets can't be used on a tubular asset, such as a stethoscope; likewise, a tag that needs a zip tie to mount can't be attached to a box-shaped asset that has no holes.

Recognizing the importance of a tag battery

Knowing whether the tag uses battery power, the exact use of the battery, and the type and life of the battery plays a crucial role in determining the cost (and hidden costs) of the tag. A tag battery sometimes also plays a role in range and real-time tag locating.

Here are some important points to consider when you choose a tag battery:

- **How the battery is used:** If the tag uses a battery, the specific manner for which the battery is used determines whether the tag is suitable for your application. Usually, if your application needs to locate tags at rates better than once every second, the battery life, battery cost, and battery replacement process may need to be examined closely, but it depends on the technology used for the tags and locating.

✔ **Battery type:** The question to ask with respect to battery type is whether the battery is *single use* (the tag needs to be replaced when the battery is dead), replaceable, or rechargeable. If the battery is rechargeable, ask whether you can recharge it without removing it from the tag, too. If you don't have to remove the battery from the tag to charge it, maintenance is much easier because you don't have to open or close the tag. However, you may not be able to replace the battery quickly to get the tag operational immediately, or some tag circuitry may be exposed to the external world.

✔ **Battery availability:** You want to know whether a battery that's used in the tag is easy to find. Sometimes, tags require a specific battery that's available only from a specific vendor. If the battery is easily available, you can save a great deal of money in long run.

✔ **Battery life:** Battery life is a concern only for semipassive or active tags. In semipassive tags, battery life is a factor of how often the location sensors request information from the tag. With active tags, battery life is a factor of how often the tag is located. If a tag is communicating (or trying to communicate) every second, that tag will have shortened battery life compared to the one that communicates twice a day with the same battery. A tag should provide a mechanism to query its battery life so that maintenance and replacement can be scheduled.

Interpreting tag enclosure ratings

The enclosure ratings identify a tag's ability to resist external environmental conditions. These ratings define the tag's resistance to everything from dripping liquid to getting sprayed with a hose to total submersion, which are all defined by these ratings.

A number of standards exist worldwide to define the type and applicability of enclosures. In North America, National Electrical Manufacturers Association (NEMA; www.nema.org — NEMA Standards Publication 250), Underwriters Laboratories Inc. (UL — UL 50 and UL 508), and Canadian Standards Association (CSA — Standard C22.2 No. 94) are commonly recognized; and worldwide, International Electrotechnical Commission (IEC; www.iec.ch — Ingress Protection (IP) Standard 60529) standards are most prevalent.

The various rating systems provide information to help you make a more informed tag choice. Although they have similar intentions, a few differences exist among them. For example, IEC doesn't specify degrees of protection against risk of explosions or conditions, such as moisture or corrosive vapors, but NEMA does. NEMA Type enclosure ratings generally follow the same guidelines as the UL Type, but NEMA doesn't require third-party testing and leaves compliance completely up to the manufacturer. UL and CSA both require enclosure testing by qualified evaluators. UL and CSA also send site inspectors to make sure a manufacturer adheres to prescribed manufacturing methods and material specifications.

For example, when choosing a tag to track wheelchairs in a hospital, you may want to consider an IEC rating IP54 or higher. An IP54 rating indicates the tag is protected against dust (no harmful deposits) and against jets of water from all directions (not immersion though).

For a detailed explanation of IP ratings and NEMA ratings, go to www.nema.org.

Tag environmental capabilities

Environmental capabilities define ranges of intended "real world" environmental conditions, such as temperature and humidity, that tags can withstand:

- ✔ **Operating and storage temperature:** Operating and storage temperature define the temperature range under which the tag is expected to operate. For example, if a tag is capable of operating in –40°C to 45°C, don't use it in places where temperatures are expected to be lower than –40°C or higher than 45°C.

- ✔ **Humidity:** Humidity defines the range under which the tag is expected to operate. So for a range of 10 percent to 90 percent, don't use the tag when the humidity is expected to sink lower than 10 percent or rise higher than 90 percent.

- ✔ **Vibration:** The source of tag vibration can be ground-borne or airborne. *Ground-borne vibration* refers to the vibration a tag receives when an asset or the surface that the asset rests on vibrates. *Airborne vibration* refers to the vibration received by the tag because of air overpressure received from the atmosphere in the form of pressure waves.

  The ability of a tag to operate normally while experiencing vibration can be a significant need for your application. For example, consider the RTLS application where a tag is attached to a machine that vibrates so that the tag can generate an alert when the vibration reaches a certain level to prevent complete machine failure.

- ✔ **Shock:** Shock tests are used to measure the fragility of products and to evaluate how they respond when subjected to a particular shock input. Shock test data for tags is key information necessary to ensure that they can withstand the intended real-world use.

  A tag's ability to handle shock is highly important for some applications. For example, if people are expected to carry tags, tags should be able to withstand being dropped repeatedly on the floor from a height of 7 feet.

  In addition to being able to withstand shock, tags can be used to detect it. Shock detection is also useful in applications, such as shipping, in which a timestamp can be recorded when a tagged asset is dropped and possibly damaged.

Supporting telemetry

Telemetry is the capability of a tag to sense and monitor the environmental conditions surrounding the tag or tagged item. The following list describes just a few of the many applications or environment-specific telemetry capabilities that enable many more applications with the same tags:

- **Temperature:** The ability to sense the temperature of tagged assets is critical for applications, such as cold storage. When the temperature goes out of range, this allows action to be taken before spoilage occurs.

- **Motion:** Tags can be used to detect when an object is moving, dubbed *motion detection*. Some businesses use motion detection to verify whether a security guard is awake or a senior is stationary in her home for unexpected durations of time.

- **Humidity:** Monitoring humidity is important for many applications, including proper food storage. Different foods require different conditions. Using tags that have humidity sensors along with temperature sensors helps ensure optimal temperatures and humidity; or at a minimum, using such tags alerts you later if the desired ranges were breached.

- **Chemicals, gases, or radiation:** Tags with chemical, gases, or radiation sensors enable people working in industrial and potentially hazardous environments to take appropriate actions to ensure safety at the plant. For example, in refineries, the ability of tags to sense gases such as hydrogen sulfide (H_2S) and carbon monoxide (CO) is desired.

- **Pesticide detection:** Tags with pesticide detection can be used on aquatic life, such as sharks, to locate them and detect pesticides and other water contaminants.

- **Bio sensors:** Tags with bio sensors, such as heart rate, breathing rate, skin temperature, and posture inclination reporting can be used to not only locate but also to monitor physiological data on the wearer, eliminating the need for multiple devices.

Connecting tags to assets

Usually tags are affixed on the assets; however, some tags can be connected physically with the asset by using wires or connectors, such as a serial cable. This setup works best for applications that require a report of the live status, or even just information, of the asset to which the tag is attached. For example, when tags are attached to IVF pumps, it may be useful to report the medicine administered, the amount of medicine present, and the status of administering the medicine. Figure 2-2 shows a tag that's connected to an IVF pump.

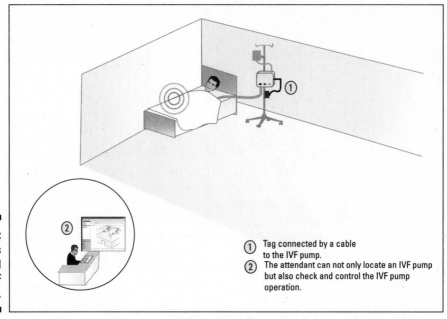

Figure 2-2:
A tag that's
connected
to an IVF
pump.

① Tag connected by a cable
to the IVF pump.
② The attendant can not only locate an IVF pump
but also check and control the IVF pump
operation.

This setup also enables applications that require remotely controlling the
asset. For example, not only can the IVF pump be located, but it also can be
remotely turned on or off.

Programming tags

To enable different configuration parameters, such as how often a tag should
transmit and what encryption schemes a tag should use in the transmitted
data, tags need to be programmed. Usually all tags are programmable in the
field, but read-only tags are programmed once at the factory or the initial
setup and can't be programmed ever again.

Tag programming refers to how the tag can be programmed, what application-
specific data it contains, and how much application-specific data can be
stored on the tag:

✓ **Program over the air or with cables:** As with any wireless application,
the capability to write and program the tag over the air, or *wirelessly,*
makes the tag easier to use, install, and update rather than accomplishing
the same tasks when cables have to be connected between the tag and
the device that's used to program it.

✔ **Application-specific data:** Some tags may have additional read-write memory that can be used to store application-specific data. Storing data on a tag can enable new applications, such as storing people's names that have accessed the tag to determine asset-use history, storing name or identification of locations a tag has visited, storing work-in-process data for a specific work order, and so on. Another parameter of interest along with read-write memory is the size of the tag's memory. The memory size indicates the amount of data that can be stored on the tag. Based on the application, this can be of huge value. For example, for a tag that records temperature every 10 seconds, if the tag memory is just enough to store 60 values, the maximum the history tag can provide is 10 minutes; however, if the tag has enough memory to store 64,000 values, the tag history can be as long as one week.

✔ **Size of the tag ID:** The tag ID size is another important consideration. The tag ID size limits the number of tags that can coexist in a facility or when the tags move from one facility to another facility. In the networking world, the tag ID is usually MAC (a 6-byte address of the networking chip that's unique worldwide), but the tag ID can be any number.

Increasing tag security

The security of your RTLS tags is important. Some tags can alert you if the tag is removed from an asset. Tag security also involves preventing hackers from opening the tag and taking data from it that compromises the RTLS or the network.

Here are some tag security features that you may want to consider:

✔ **Tamper detection:** The tamper detect feature enables a tag to detect its unauthorized removal (or *tampering*) from a person or an asset. This capability is critical for many applications, such as asset theft prevention in which the whole purpose of tags is defeated if the tag can be removed without alerting, or the detection of the unauthorized tag removal used in an electronic house arrest monitoring system.

✔ **Hack proof:** The hack proof feature refers to an unauthorized person, even with physical access to the tag, who can't read any data from the tag memory that can compromise the RTLS system or the network. One way to think of this is that all keys and important data aren't stored in the nonvolatile memory and/or all stored data is encrypted. For example, in a Wi-Fi RTLS, it may be required for all associating tags to comply with Payment Card Industry (PCI) standards that support Wi-Fi Protected Access (WPA2).

Working on tag serviceability

Tag serviceability means that you can service a tag or do maintenance work. More specifically this has two aspects:

- ✓ **Software serviceability:** You can perform tasks, such as downloading engineering-usable data from tags and updating *firmware* (software inside the tag) remotely to address any field bugs; you don't have to move the assets or tags to the vendor site to troubleshoot problems or update security patches.

- ✓ **Physical serviceability:** Implies the ability to clean tags under cleaning conditions of the corresponding asset. If tags are attached to hospital equipment, for example, they can withstand the cleaning, sterilizing, and disinfecting procedures used for the corresponding equipment.

Making Sense of Location Sensors

Location sensors are devices within an RTLS that typically have a known position and detect the location of tags. Location sensors locate tags by using a *physical parameter,* or a measurement, that exists between the sensors and the tags. The physical parameter can be something as simple as *visibility* — the tag is in the line of sight of the sensor — or it can be more complex, such as measuring the time a signal takes to travel from the tag to the sensor.

Depending on the technology, these sensors can be some devices that leverage the infrastructure already present in the facility, or they can be a brand new infrastructure added to the facility (also called as an *overlay*). For example, in one technology where Wi-Fi is used for the RTLS in an enterprise, existing Wi-Fi access points (that are deployed to provide Wi-Fi data network in the enterprise) may be used as the location sensors for the RTLS. Or in another technology that uses ultrasound for an RTLS in a hospital, microphone receivers in existing computers (that are deployed in every room to provide patient data access to caregivers) may be used.

The number of location sensors needed in a facility usually depends on the technology, the application, and the desired accuracy. For example, if you're implementing an RTLS for a wandering management application (where you want to record and monitor whenever a senior leaves the facility), you might need location sensors only at all the doors (exits) and implement the RTLS only by choke points. In another example, if you are trying to implement an RTLS to do an evacuation management application where you just want to know if an employee is in the building using a Wi-Fi RTLS and you already have Wi-Fi deployed in the building, you may not need any additional sensors.

Inspecting the Location Engine

The *location engine* is the software that communicates with tags and/or location sensors to determine the location of tags and report it to middleware and/or applications. Middleware and applications are covered later in this chapter, in the sections, "Understanding Middleware" and "Understanding RTLS Applications."

Location determination basically consists of

- ✔ **Ranging techniques to estimate distance between the tag and a set of location sensors:** Do this by using some physical variable, such as the time of flight of the signal from the location sensor to the tag. Various techniques are covered in the next section, "Ranging techniques."

- ✔ **Position estimation techniques that derive the position of the tag:** This includes making use of *estimation algorithms* (covered in the section, "Exploring the position estimation algorithm"), such as triangulation, on all the estimated tag distances from the location sensors and the actual position of all the location sensors to the estimated tag position.

Ranging techniques

Ranging techniques estimate distance, or range (usually in terms of feet or meters), between the tag and the location sensor. Here are just a few of the physical variables you can use to determine estimated distance:

- ✔ **Proximity:** Proximity measures the nearness to a known set of points. The exact physical variable used for proximity is different for each technology, and the estimated distance depends on the technology. For example, in a passive RFID network, the antenna location itself localizes the tag within the antenna read range (say 20 feet); in an 802.11 network, the tag's ability to associate can be used for proximity sensing (say 150 feet in indoor environments); and, for closed-circuit television-based systems, the image captured determines proximity.

- ✔ **Time of Arrival (TOA):** TOA basically makes use of the time it takes for a signal to travel from the location sensors to the tag or vice versa. This time to travel, also known as *propagation delay,* can be converted into distance between the tag and the location sensor by multiplying it by the signal's *propagation speed* (the speed at which the signal propagates, or disperses, into space). At high frequency, such as 2.4 GHz, the signal travels at a speed approaching the speed of light. If the signal takes 0.1 microseconds to travel from the location sensor to the tag, it indicates a distance of 30 meters between the tag and the location sensor.

✔ **Angle of Arrival (AOA):** The angle between the propagation direction of signal and some reference direction, which is known as orientation. The AOA uses direction-sensitive antennas by the receivers (the location sensors) to determine the direction (and the angle) of a signal from the tag to the location sensor, and the tag position is estimated by finding the intersection of different signal propagation paths. AOA is highly range dependent and a small error in the angle measurements will result in a large location error if the location sensor is far away from the tag. Also, not having a line of sight between the tag and location sensor is a significant error source in AOA-based locationing.

✔ **Time Difference of Arrival (TDOA):** The TDOA is very similar to TOA. Instead of the exact TOA measurement, which requires high-resolution synchronized clocks at the tag and the location sensor, TDOA measures the difference in transmission times between signals received from each of the transmitters to a tag or vice versa. That TDOA is used to estimate distance between the tag and the location sensors. Although the distance between the tag and the location sensor doesn't matter, it is required that for TDOA measurements, all location sensors must be exactly time synchronized. Otherwise, location ambiguity will be introduced.

✔ **Time of Flight (TOF):** The TOF method uses measured elapsed time for a transmission between a tag and a location sensor based on the estimated propagation speed of a typical signal. The signals are sent with known departure times to tags, so the tag knows the flight time.

✔ **Round Trip Time (RTT):** This method uses the total time for a signal to start from the location sensor and the acknowledgement to be received back.

✔ **Received Signal Strength Indicator (RSSI):** This is a measurement of the power present in a received radio signal. As a signal leaves its source, it attenuates, meaning that the power of the signal drops. The drop is logarithmic, and the signal attenuation in open space, as well as through different mediums, is well defined. Because the power levels at the start of transmission of a signal are known, RSSI can be used to estimate the distance the signal has traveled. Using RSSI sounds easy, but RSSI poses an interesting challenge because it is affected by obstacles, multipath fading (when an RF signal takes different paths when propagating from a source, such as from a tag to a location sensor), temperature and humidity variations, opening and closing of doors, furniture relocations, the presence and mobility of human beings, and so on. What this means is that the value of RSSI as seen by the receiver could be much higher or lower than the theoretically anticipated value.

In general, no single variable can be used to provide accurate ranging estimation under all circumstances. Each variable has its own advantages and limitations in terms of location accuracy, and many times, a combination of some of the preceding variables is used.

Exploring the position estimation algorithm

For given ranging techniques, several *algorithms* (mathematical formulas) can be used to compute the position. Here are just a few of those algorithms:

✔ **Trilateration:** Trilateration is a technique in which you can estimate the position of something if you know its distance to three different locations.

When you use more than three locations, it's called as *multilateration*.

For example, imagine you're totally lost — for whatever reason, you have absolutely no clue where you are. Then someone tells you that you're 45 miles from San Francisco, California. This helps a bit because it means you're somewhere on the perimeter of a circle with a 45-mile radius with San Francisco at its center. If you also know that you're 120 miles from Sacramento, California, you know that you're at one of the two intersecting points of the two circles — one with the 45-mile radius with San Francisco at the center, and the other with the 120-mile radius with Sacramento as the center. If you also know that you're 345 miles from Los Angeles, you can be in only one place — Santa Clara, California, where these three circles intersect. With trilateration, you can compute the location precisely. Figure 2-3 shows this example of trilateration.

✔ **Triangulation:** Triangulation is a technique in which you can estimate the position of something if you know the line angle between that something and the three different locations with respect to a common reference line, such as a line pointing up.

For example, imagine that you're sitting on the ground but don't know where — for whatever reason. Someone on the top of a tower, say Tower A, says he sees you at a 135-degree angle from a line pointing up. This means you can be anywhere on a slope at an angle of 135 degrees from that point and that slope can be anywhere around the tower (geometrically, a conical shape). Now, someone else says she sees you at a 160-degree angle from the peak of another tower, Tower B. You're getting closer because now you can be at only the intersecting points of those two conical shapes (the slope intersections from the peaks of the two towers). When someone else says you're 135 degrees from the peak of a third tower, Tower C, you can be at only one position, the point where the three angular lines intersect. (See Figure 2-4.) One can compute exact location when the angles to three positions are given.

People often use the *triangulation* when they actually mean *trilateration*. Don't let them confuse you. Ask for clarification.

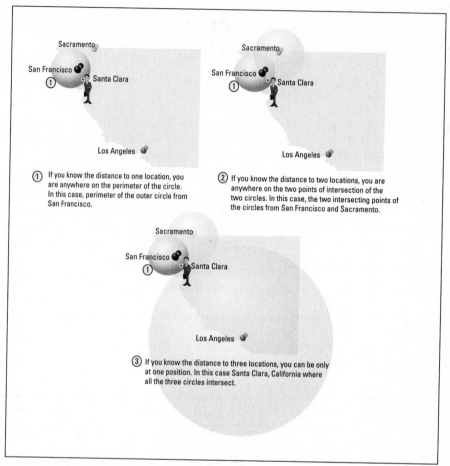

① If you know the distance to one location, you are anywhere on the perimeter of the circle. In this case, perimeter of the outer circle from San Francisco.

② If you know the distance to two locations, you are anywhere on the two points of intersection of the two circles. In this case, the two intersecting points of the circles from San Francisco and Sacramento.

③ If you know the distance to three locations, you can be only at one position. In this case Santa Clara, California where all the three circles intersect.

Figure 2-3:
Trilateration
helps
estimate
positions.

✔ **Scene analysis:** Scene analysis examines a view from a particular vantage point to draw conclusions about the observer's location. The scene itself can contain visual images, such as frames captured by a wearable camera or any other measurable physical phenomenon, such as electromagnetic characteristics that occur when an object is at a particular position and orientation. If you can see the swimming pool from the hotel room, for example, you must be in the east wing and on the 8th or 9th floor. Compared to the preceding two approaches, scene analysis requires much more prior knowledge about the environment.

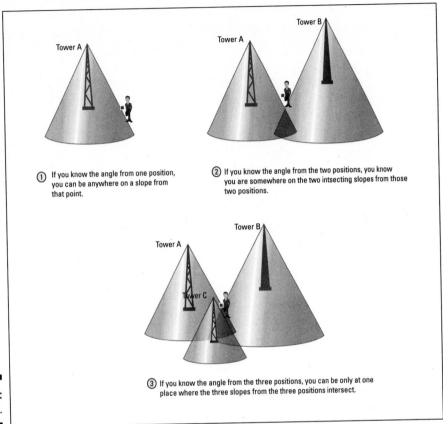

① If you know the angle from one position, you can be anywhere on a slope from that point.

② If you know the angle from the two positions, you know you are somewhere on the two intsecting slopes from those two positions.

③ If you know the angle from the three positions, you can be only at one place where the three slopes from the three positions intersect.

Figure 2-4: Triangulation.

✔ **Nearest neighbor:** In this technique, instead of applying a finite algorithm, simple neighbor relationships are used to estimate a position. A *neighbor relationship* is based on any of the ranging techniques, such as RSSI (see the preceding section, "Ranging techniques"). If the RSSI of a location sensor as seen by a tag is greater than –50 dB, you can assume the tag is in the neighborhood of that sensor. For example, as shown in Figure 2-5, if A can detect that you're in the neighborhood and B and C can't detect, you're in the smaller A zone. Or, if A and B can detect your presence but C can't, you must be in the zone between A and B, away from C. With strategically placed neighbors (location sensors), tag location can be determined. Figure 2-5 illustrates the nearest neighbor technique.

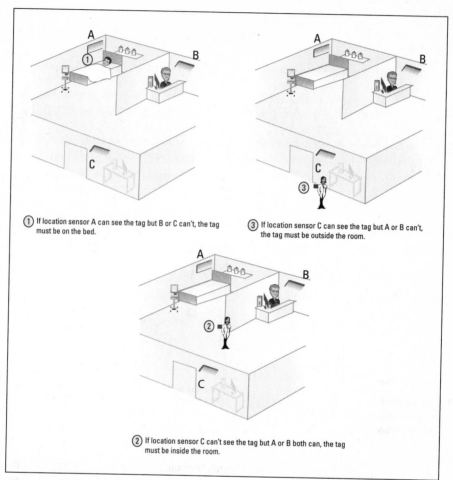

① If location sensor A can see the tag but B or C can't, the tag must be on the bed.

③ If location sensor C can see the tag but A or B can't, the tag must be outside the room.

② If location sensor C can't see the tag but A or B both can, the tag must be inside the room.

Figure 2-5:
Using the nearest neighbor technique.

Applying the earlier techniques to compute tag locations

The location engine makes use of one or more ranging techniques and position-estimating algorithms, as described earlier, to compute the tag locations. The following sections examine several examples.

Using the Angle of Arrival method

Each location sensor computes the Angle of Arrival (AOA) of the signal received from the same tag in reference to a fixed direction, say north, and forwards the value of this angle to the location engine. The location engine uses the position of location sensors along with the value of angles to construct the signal path from the tag to each location sensor, and the location of the tag is nothing but the intersection of these two signal paths.

You can determine position of a tag in a two-dimensional (2D) plane using just two location sensors and using three location sensors to compute position in a three-dimensional (3D) plane.

Figure 2-6 illustrates this method. In the figure, two location sensors, #1 and #2, are used and tag T's position is being determined.

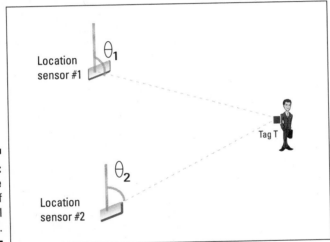

Figure 2-6:
Using the
Angle of
Arrival
method.

One common approach to obtaining AOA measurements is to use an antenna array on each location sensor. The accuracy of this method increases with the number of antenna arrays used in each sensor, and it also increases cost. In addition to the cost, the resulting angle measurements are rather sensitive. A small error in the angle measurements will result in a large location error if the location sensor is far away from the tag. The AOA method is best suited for direct line-of-sight measurements between tags and location sensors.

The AOA method is susceptible to security threats because attackers can easily reflect or retransmit from a different location.

Using the Time of Arrival method

Each location sensor notes the time of arrival of signal transmitted by the tag (say started at time moment t0) and computes the elapsed time (expressed as time of arrival of signal at the location sensor — t0). It then forwards this elapsed time to the location engine. The location engine uses the position of location sensors along with the elapsed time (which effectively identifies the distance between the tag and the location sensor) and determines the location of a tag by using *trilateration,* which is viewed as an intersection of circles (in 2D space) or spheres (in 3D space), with location sensors as the center and the distance as the radius. See Figure 2-7.

You must have at least three location sensors in a two-dimensional (2D) plane and at least four location sensors in a three-dimensional (3D) plane.

At high frequency, such as 2.4 GHz, the signal travels at a speed approaching the speed of light (signal can travel 30 meters in 100 nanoseconds). And, hence, to achieve precision up to the nanosecond scale, high-precision clocks in tags and location sensors, as well as an elaborate clock synchronization system (so that all location sensors and tags refer to identical time moments such as t0) must be used.

More accurate clocks provide better accuracy, but at a higher cost to the system. The need to have at least three location sensors in both 2D plane and 3D space adds to the cost and complexity as well. Figure 2-8 illustrates the location of a tag in a 3D plane.

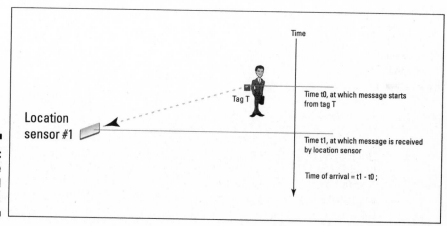

Figure 2-7:
The Time
of Arrival
method.

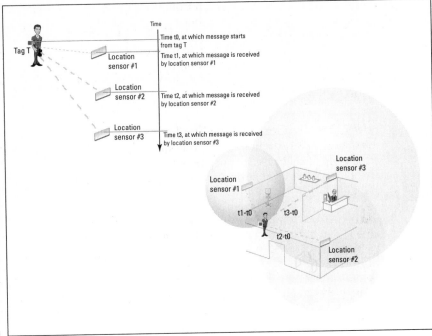

Figure 2-8:
Using the
Time of
Arrival
method
to locate
in three
dimensions.

Using the Time Distance of Arrival method

The Time Distance of Arrival method is similar to the TOA method except for the fact that the start time of signal from tag (the time moment t0 as used in the TOA method) isn't used. The TDOA method measures the *difference* in transmission times among signals received from each of the tags to the location sensors.

Each location sensor notes the time of arrival of signal transmitted by the tag and forwards this time to the location engine. The location engine uses the position of location sensors along with the received signal's time difference between each of the location sensors. Mathematically, for trilateration purposes, tag position computed by AOA is at the intersection of spheres with center points of location sensors, and the tag position computed by TDOA is the intersection of hyperbolas (hyperboloids in 3D).

Like TOA, three or four location sensors are required at known fixed positions for TDOA.

The problems related to the TDOA method are similar to those of the TOA method. The precision of the location engine is correlated to the accuracy of the clocks. TDOA requires high-precision clocks in all location sensors, as well as an elaborate clock synchronization system. However, unlike TOA, the clock of a tag doesn't need to be synchronized.

Many times, signals don't reach from the location sensor to the tag (or vice versa) in a straight line but rather by bouncing from various surfaces in the environment; the time difference, which is already a small number, is affected adversely by the longer paths taken by some signals. Inaccuracy in arrival time differences results in inaccurate computations based on TDOA. For TDOA methods, the direct line of site is preferable, such as in an open space or in large open buildings.

Using the Time of Flight method

The Time of Flight (TOF) method is similar to the TOA method in the sense that it uses measured elapsed time for a transmission between a tag and a location sensor, which the location engine uses to compute location by trilateration.

However, unlike TOA, the signal is transmitted from the location sensors to the tag and the time of start of the signal is embedded in the signal itself (or is well known). Because the time of start of the signal is embedded in the signal itself or well known, the system doesn't require an elaborate clock synchronization system. However, because this method is based on a time value, the quality of the clock becomes significantly more important than in the TOA or TDOA methods. The clock offset and clock drift can significantly reduce location accuracy.

Like TOA or TDOA, three or four location sensors are required at known fixed positions for TOF. Figure 2-9 illustrates the TOF method.

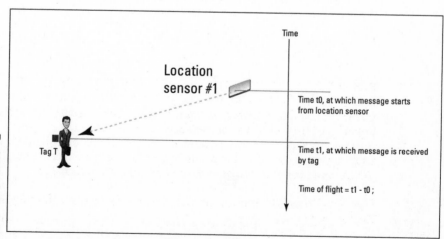

Figure 2-9:
Using
the Time
of Flight
method.

And like TDOA, if the signal bounces off various surfaces with no direct line of sight for the signal to travel from the location sensor to the tag (or vice versa), the flight time can be inaccurate. Small inaccuracies can result in large errors in location precision (because the signals travel at a speed approaching the speed of light, an error of 0.1 nanoseconds is equal to an error of 30 meters).

TOF has an advantage over other time-based systems because no elaborate clock synchronization hardware or cabling is needed.

Using the Received Signal Strength Indicator method

In this method, the location engine makes use of RSSI of the signal as received by a tag from various location sensors, or alternatively, RSSI of the signal from a tag as seen by all location sensors, to estimate the position of the tag. The tag location is computed by one of the following methods:

- ✔ **Nearest neighbor:** In this technique, the tag is assumed to be near the location sensor that receives the largest RSSI value.

- ✔ **Scene analysis:** RSSI is used as the physical phenomenon for scene analysis. A database of possible RSSI values from various location sensors in different parts of the facility is built beforehand. This is accomplished by visiting the site with a real tag and doing a walk-through of the facility (this process is a *site survey*). And in real time, when the tag reports the RSSI values from various location sensors to the location engine, the location engine does a best (or closest) match of the values with the database and returns the corresponding location.

  The biggest disadvantage of this method is that the RF environment is usually dynamic and the RF conditions during which the database was made may not match the current RF conditions. Not only do you need to build this database beforehand, but you also need to update this database periodically. However, you might find tools to build and update this database automatically.

- ✔ **Trilateration:** The RSSI of signals, as shown by the receiver (as shown by the tag for signals from the location sensor, or as shown by the location sensor for signals from the tag) is used to estimate the distance between the tag and the location sensor. Consider an example in which the tag determines the RSSI of signals as received from the location sensor (the mechanics of the model in which the location sensor determines the RSSI of signals from the tag are very similar):

  • Each location sensor transmits signals at a preconfigured power.

  • The RF signals emitted by a location sensor go through significant attenuation, even in free space (such as no obstructions between the transmitter and the receiver), before they reach the intended recipient. The propagation loss is typically given by the following formula:

Path loss in dB = $C + 10 \times n \times \log_{10}$ (d)

Where *n* is the path loss exponent, *d* is the distance between the transmitter and the receiver, and *C* is a constant which accounts for systems. The value of *n* depends upon the building type such as vacuum, retail store, grocery store, office with hard partitions, office with cubicles, industrial, and so on.

- By knowing the propagation loss (such as the starting transmit power of the signal and the RSSI as shown by the tag), the distance between the tag and the location sensor can be computed. And, as described in the section, "Exploring the position estimation algorithm," trilateration (using the distance between the tag and each location sensor) can be done to compute the position of tag.

Because RSSI changes exponentially with distance, an RTLS with RSSI usually requires a dense deployment of location sensors if high accuracy is desired. This may add considerably to the system's cost. However, the key problem related to RSSI-based systems is that in practice, a radio signal may encounter many objects in its transmission path and may undergo additional attenuation, depending on the absorption characteristics of the objects. There are many types of objects, including fixed (such as concrete walls or glass windows), mobile (such as carts), and transient objects (such as people) that absorb RF energy and cause RF attenuation. An adequate underlying path-loss model must be found for such conditions as because of this, actual estimated distances are somewhat unreliable.

Using Round Trip Time

Different methods that make use of TOF or TOA need high quality clocks and optionally highly synchronized clocks between tags, location sensors, or both.

In Round Trip Time-based methods, a signal is sent from the location sensor to the tag, and an acknowledgement is sent by the tag back to the location sensor. The location sensor makes use of this Round Trip Time (RTT) of a signal to determine the distance between the tag and the location sensor, and then the location engine computes the tag location with trilateration (using the distance of the tag to three or more location sensors and the location of the location sensors).

If the tag can acknowledge the signal back in a highly predictable manner, RTT methods can alleviate the need for highly expensive synchronized clocks and can be used very effectively to compute position. See Figure 2-10.

Note that the RTT method can be implemented in reverse order in which the RTT is computed by the tag by sending a signal to the location sensor and then the location sensor sends an acknowledgment back to the tag. Optionally, the timestamp measurements of RTT can be taken by both the tag and the location sensor to provide two measurements of RTT that can then be averaged.

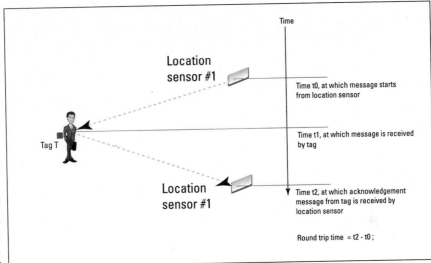

Figure 2-10:
Using the
RTT method.

Exploring the Technologies

Many technologies are available or being developed that enable an RTLS with one of the variables and the algorithms described in the previous sections. Here are just a few (note that most of these technologies and how they're used in an RTLS are detailed later in this book):

- Bluetooth
- Building illumination
- Digital TV signal
- GPS
- Infrared
- ISO 24730-2
- Passive RFID
- Powerline positioning
- RuBee
- SAW
- Ultrasound
- Ultra Wideband
- Wi-Fi
- ZigBee

Each of these positioning technologies has a certain level of accuracy and a service area where it performs better. Although they've been introduced individually over time by different organizations, universities, and vendors, their value is readily apparent when they're considered together, as complementary innovations.

You can't rely on any one technology to provide accurate location information in all environments. Each technology has its pros and cons, as I discuss throughout this book.

Understanding Middleware

Middleware is a generic term used to describe software that connects two disparate applications, allowing them to communicate with each other and to exchange data. In the context of an RTLS, middleware is the software that resides among the pure RTLS technology components (tags, sensors, and the location engine) and the business applications.

The key objective of middleware is to make the applications independent of tag (or RTLS) technology. Middleware functionality includes

- ✔ **Tag, location sensor, and location engine management:** Middleware enables applications to configure, monitor, deploy, and issue commands directly to the location sensors and tags through a common interface. For example, a command to know if all location sensors in a specific physical area are operational or not; a command to start, stop, or restart location engine; and so on.

- ✔ **Location data management:** Middleware does filtering and routing of location data to the appropriate destinations (applications).

  - *Filters* by removing duplicate events or business rules as desired by applications

  - *Consolidates* when multiple events are received for the same tag, for example

  - *Routes* the events to the appropriate application

  - *Stores* the events and maintains history

- ✔ **Application integration:** Middleware provides the messaging, routing, and connectivity features required to reliably integrate data into different existing applications of the enterprise.

- ✔ **Business rules and process management:** Optionally, middleware also processes and manages business rules that touch many applications. For example, a business rule that knows when a tag moves from one zone to another can be used by an asset security application, a security guard monitoring application, and so on.

✓ **Architecture scalability and administration:** Middleware also enables advanced features, such as dynamically balancing processing loads across multiple servers, automatically rerouting data upon server failure, and so on to enable an RTLS architecture that's scalable and an enterprise class.

Figure 2-11 shows a typical RTLS middleware architecture.

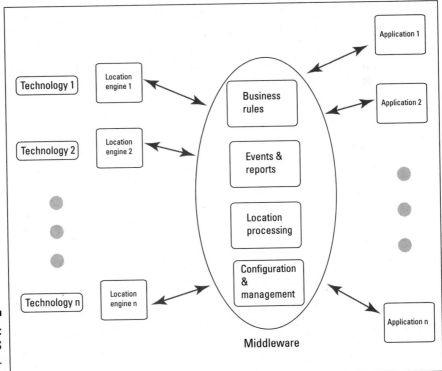

Figure 2-11:
RTLS
middleware.

Understanding RTLS Applications

In an RTLS, an *application* is the computer software that interacts with the RTLS middleware to solve the problems challenging end users, such as enabling users to achieve the tasks that they wish to perform or, in some cases, solves problems for another application program. For example, in an application that enables asset security, the application may constitute software that

✔ Provides a user interface that allows a user to configure what constitutes an alert, such as if an asset leaves the premises without any authorized user around, if an asset leaves the premises after 6 p.m., and so on.

✔ Raises alerts based on user configuration.

✔ Shows current and historical alerts (as well as alert locations).

✔ Sends the alerts as e-mails, SMS messages, messages to pagers, or pre-recorded voice messages to security.

✔ Takes care of appropriate escalations if the alert isn't cleared in a specific period of time.

The application provides the end user value — the real-time actionable business intelligence that's accessible to the relevant systems and people. It involves the establishment of alerts, alarms, actions, decisions, audit trails, and documentation.

Chapter 3

Choosing the Right RTLS

. . . .

In This Chapter

▶ Knowing your options

▶ Figuring out your objectives and deployment needs

▶ Estimating maintenance and integration needs

▶ Measuring Return on Investment and value adds

▶ Evaluating your security

▶ Tracking performance and requiring a standards-based solution

▶ Becoming eco-friendly and making your decision

. . . .

You may implement an RTLS because you have a problem that you want to solve. Perhaps you have some expensive equipment that's getting lost or stolen and you want to know whether that equipment leaves your building. Or maybe you need to locate a doctor or nurse to determine how much time the attending physician will take to reach a patient.

To find an RTLS that fits your needs, you must compare competing technologies, hardware, solutions, and associated vendors. However, the comparison's outcome on its own may not tell the full story because not only are there certain ancillary needs that you need to consider but also because you'll learn more about the RTLS and may grow to use it in new ways. For example, you may install the RTLS in one facility and later take it to others, or you may use the RTLS to track visitors and later use it for employee safety.

This chapter provides an overview of the decision-making process involved when choosing an RTLS that fits your needs.

Weighing Your Options

When managers make a business decision, they typically make a decision matrix. A *decision matrix* helps managers structure the criteria under consideration. To create a decision matrix, you specify and prioritize your needs. Then you evaluate, rate, and compare potential solutions and then select the best matching solution.

To build a decision matrix for RTLS, first identify the building blocks of an RTLS application:

- **Basic objectives:** This refers to the key objectives of the RTLS application, such as accuracy, scope, responsiveness, and life expectancy.

- **Deployment:** Installation is a significant consideration when you create a decision matrix. You must consider all the things that impede or enable the installation of the RTLS application — impact to the existing infrastructure, coexistence with existing technologies, installation wiring, ease of deployment, and more.

- **Usability:** Solutions shouldn't depend on people to do the right thing. People shouldn't need extensive training courses showing them how to use the system. This and many usability issues also often distinguish one RTLS from another.

- **Maintenance:** Maintenance issues take time and cost money. Maintenance includes tasks, such as replacing tags, expanding applications to other areas, and upgrading software.

- **Integration:** *Integration* is the ability of an RTLS to exchange information with other systems, such as employee time-card management systems, product recall management systems, warehouse management systems, material management systems, and so on. The ability of your RTLS application to use the enterprise data from other applications — and the reverse — provides end users with useful data for improved and more intelligent decision making. This may be an important criterion for distinguishing among different RTLS.

- **Return on Investment:** System cost, which includes the cost of installation, deployment, infrastructure, and maintenance, is often a crucial parameter that distinguishes one RTLS from another. If you're looking for a locating system, it's important to select a system that best meets the needs and generates a compelling Return on Investment (ROI).

- **Value adds:** This refers to all the value adds over and above the basic application that you may be able to use in the short or long term. For example, not only can you locate, but your tags also provide information about ambient temperature.

- **Security:** Due to the RTLS's wireless nature, security is a great challenge. Malicious hackers don't have to have physical access to hack or attack an RTLS. For this reason, the list of possible threats and countermeasures is an important consideration.

- **Performance tracking:** This refers to all the indicators that an RTLS has to help you measure whether you're successful, progressing in the right direction, and what changes, if any, you should make.

✔ **Standards based:** Determine whether the system is based on real industry standards or whether you could get locked in with silos.

✔ **Eco-friendliness:** A different RTLS adds different levels of hazardous or non-hazardous waste (for example, the batteries used by tags). You may want to compare the RTLS based on the levels of pollution each adds per year.

Knowing Your Objectives

First and foremost, you must know the basic objectives for your RTLS, or the goals that you wish to achieve by using it. You must know the problem that you're attempting to solve. The information in the following sections helps you define your application as clearly as possible.

Scope of the RTLS application

The *scope,* or coverage, of an RTLS carries a significant weight regarding whether an RTLS is better than another for an application. Are you trying to locate indoors or outdoors? Do all areas of the facility need coverage, such as can the asset go to every part of the facility or is the asset confined to an area? What will be the density of the location sensors that you'll need to locate tags? Each RTLS addresses these questions differently. Some RTLS can do room-level accuracy but can't locate assets that are in hallways or parking lots, and others can locate them in hallways and parking lots but can't give room-level accuracy. Some need the deployment of a large number of location sensors to cover an area, and others can cover even larger areas with fewer location sensors.

Responsiveness

The *responsiveness* of an RTLS is defined as how quickly the location system outputs the location information. And, if the tracked objects, such as people, are constantly moving, latency in the estimate translates into a less accurate outcome. An RTLS may have the best level of everything, such as accuracy, size, and battery life, but if it's not as responsive as desired by the application, it just won't work for you.

Timeliness becomes very important, especially with safety-related applications. If the system reports a location estimate with a 60-second delay, the object or person may have moved to another floor in that time, or the security personnel who came to the reported scene may miss the situation completely.

Accuracy

Accuracy measures how close — or far — an estimated position is from the true position. You express accuracy with an accuracy level and a precision value. For example, you may see accuracy expressed as 15 centimeter accuracy over 95 percent of the time. The accuracy level indicates how "far" the estimated position is from the real position. The precision indicates how often you can expect to capture a measurement at the given accuracy level (or better).

The accuracy with which an RTLS locates is often used to determine whether the chosen system is appropriate for certain applications. Different RTLS locate with different accuracy, and the accuracy is usually expressed as described in the following list:

- ✔ **Zone level:** Location engine reports whether the tag is located in a zone. If an RTLS says 95% Zone Level Accuracy, the tag is reported in the correct zone 95 percent of the time.

- ✔ **Room level:** Tag location is reported to be in a room or not. If an RTLS says 90% Room Level Accuracy, the tag is reported to be in the correct room 90 percent of the time. Usually room size is also specified along with this.

- ✔ **Sub-room level:** Tag location is reported to be in the sub-area of the room or not. If an RTLS says 90% Sub-Room Level Accuracy, the tag is reported to be in the correct sub-area in the room 90 percent of the time. Usually room size and sub-area size are also specified along with this.

- ✔ **Association level:** Tag location is reported to be next to another tag. If an RTLS says 95% Association Level Accuracy, the tag will be reported correctly next to another tag 95 percent of the time. Usually a distance for association level is also specified along with this.

- ✔ **Exit/entry level:** Location engine reports tag location as seen at a specific entry or exit point. If an RTLS says 95% Exit/Entry Level Accuracy, the system might not see entry or exits 5 percent of the time. Usually maximum tag speed at which it can travel is also specified because the accuracy is usually valid for speeds lower than this number.

- ✔ **Distance:** Tag location is reported within this distance from the real position. If an RTLS says 95% Accuracy of 12 Feet, the tag is reported to be within a 12-foot radius of the actual position correctly 95 percent of the time.

Size

Although the accuracy of an RTLS is a foundational property for modeling location-based applications, one factor that distinguishes one RTLS from another is the size of tags the RTLS require. If the assets you plan to track are tiny, you may not be able to use an RTLS that makes use of big RTLS tags even if its accuracy fits your application best. And, conversely, if the assets you're planning to track are huge, you may not want to use teeny-weeny tags. For example, if you want to attach tags to jewelry, the tags can't be big; and when you're trying to put tags on shipping containers, you'd rather have a tag big enough so that you can find the tag on the container.

In some applications, you may want a big, visible tag because just the tag size can deter or discourage a thief; examples include infant abduction prevention applications and laptop theft prevention applications. In other applications, such as shopping cart theft deterrent systems, size isn't as important because theft deterrence is done by making the RTLS tag part of an electronic-mechanical system. The system locks one of the wheels, usually one on the front, when the cart is rolled out of a designated area.

Life expectancy

How long a tag lasts, how often the tag or its battery has to be replaced, and whether the tag gives an indication that its battery life is about to expire (or just dies abruptly) is vital information for any application. The tag life is usually a technology factor that's used by the RTLS. An RTLS may have the right accuracy and the right tag size, but you may not be able to select that RTLS if the tag's life doesn't match the application's demands. For example, if you're tagging something that's going into a submarine, you want tags that last long enough and don't need constant replacement. On the other hand, if you're using tags as a day badge for visitors, the tag battery life doesn't have to be more than a day.

Tolerance

Tolerance refers to the ability of the RTLS to produce good and consistent results. This takes almost the same priority as the accuracy of the RTLS. In environments that change a lot, such as the number of people present or some temporary obstacles moving on or out, an RTLS with a high tolerance for changes in the environment is better than the RTLS that locates erratically with the changes.

Scalability

Scalability refers to the maximum number of tags that can be located by the RTLS as well as the number of individual tags that can be located simultaneously.

As your facility or operation grows, you'll need to track more and more objects, which puts pressure on the RTLS's scalability and computing speed. The facility's size may also generate problems if the RTLS hasn't been designed to scale to larger location areas.

The ability to balance tags across multiple servers should be a prerequisite for growing a system that must track hundreds of thousands of assets.

Assessing Deployment Needs

Deployment, or installation, is a one-time effort. Well, sort of. You install the system initially and then reinstall as you expand it. Even so, deployment presents a huge consideration. You need to consider the cost, effort, and time commitment required to achieve it. The following sections cover some other important considerations related to deployment, as well.

Anything that's difficult to install is usually difficult to remove, which means you might get stuck with an RTLS technology or a vendor for a long, long time. So plan carefully and choose wisely.

Environmental suitability

Some RTLS technologies work better in one environment than another. The environmental considerations include the physical layout of the area where RTLS is needed, the amount of metal, water, and gases in the environment, how frequently an environment changes, existing technologies, and more.

Impact to an existing infrastructure

The impact to an existing infrastructure is often a huge consideration because the technology used by an RTLS could interfere with the technologies that already exist in your facility. An RTLS that can potentially interfere with existing technologies isn't good for either the RTLS itself or the products that are using those technologies.

Leverage of an existing infrastructure

Many technologies are already deployed in businesses today. If everything else is equal, an RTLS that can leverage the existing infrastructure could be of enormous value as it can help ease the deployment and maintenance.

Ease of deployment

Each RTLS gets deployed in its own way, and the ease of deployment of one RTLS over another with respect to time, effort, and cost can contribute significantly toward your selection of a particular RTLS.

The questions to ask regarding the ease of deployment include

- Can you implement the system with minimal disruption to your operations?

- Does the RTLS require low support and management as well as minimal training for staff?

- Does the solution fit into your way of doing things to facilitate adoption by your personnel into their workflow?

- Is the system mature to the point that the tools are intuitive and easy to learn and use?

Installation effort

When considering deployment of location sensors, different RTLS technologies have different requirements with respect to

- **Cabling installation needs:** Many RTLS technologies require that location sensors are *hardwired* (holes are drilled in walls or ceilings) or tiles from dropped ceilings are removed and replaced in order to run cable for each location sensor.

  Cabling and wiring requirements may not be an important consideration when construction is ongoing or when the walls are exposed. However, new cabling requirements can have a huge impact when your facility is already in place and operational. Think about the following:

  - *Dynamic facility:* If your facility has areas that are sometimes rearranged for new equipment, moving furniture or aisles, or department expansions, the cabling and repositioning of location sensors can add substantial effort and cost.

- *Mobile equipment:* If you have equipment that sometimes goes to temporary facilities (such being put in tents during an emergency), it may be impractical to do wiring/cabling in those environments, which limits your ability to locate in those environments.

- *Facility-wide coverage needs:* To maximize the benefits of RTLS deployment, you may want to cover all parts of your facility, whether they're indoors or outdoors. For example, in a hospital, without facility-wide coverage (buildings, open spaces, and parking garages), you simply can't enable emergency response for your caregiver safety or asset theft detection.

Cabling/wiring needs can affect your ability to evaluate or validate the RTLS technology or application effectively. Cabling/wiring can be expensive, labor intensive, time-consuming, and disruptive to your operations. Cabling/wiring needs aren't just an installation challenge; these can also result in additional maintenance overhead.

✔ **Sensor orientation:** Many RTLS technologies require that the location sensors be placed in a specific orientation, at a specific height, or at exact laser-verified coordinates in the building.

This can be an important challenge because this necessitates a specialized installation crew. The skill sets and competencies needed for installation can impact not only the installation itself but can also force you to get locked in with a specific vendor or provider for installation.

✔ **Sensor power needs:** Typically RTLS technologies require that location sensors be powered electrically. If the RTLS solution you're deploying depends on power outlets, consider the following:

- *Outlet availability:* With electrical outlets already at a premium in many facilities, such as hospitals, will you have open outlets at your disposal to plug in the sensors? What if you need your RTLS solution to work in an area where no physical outlet is present? How would you locate equipment or staff outdoors? Your solution will be limited to where the outlets are already in place, or you'll have to install new outlets at the desired location.

- *Outlet accessibility:* With existing outlets in high demand, consider who has access to the outlets. Is it possible for someone with malicious intentions to remove the location sensors? Is it possible for the cleaning crew or the maintenance staff to unplug the location sensor and use that outlet for their immediate need, such as to use a vacuum or power tools? Is it possible for anyone to unplug the location sensor by mistake?

- *Outlet locations:* Usually outlets are installed within a couple feet of the floor. With sensors placed near the floor, the RTLS technology may not perform optimally. Does the RTLS solution necessitate extension cords to bring location sensors to a specific height for optimal results?

- *Wall warts: Wall wart* is a slang term for over-sized AC plugs that are typically packaged with electrical devices that don't contain their own power supply. There are many problems in using wall warts. For instance, wall warts may obscure the other socket in an outlet; people tend to remove a plug to put in another; the weight of a wall wart may cause it to fall out of the power socket; and a typical wall wart usually draws power even when the attached device isn't in use, resulting in power consumption inefficiency.

- *Dedicated circuit needs:* Having significant power outlets consumed for a specific application's location sensors may necessitate the need for dedicated circuit breakers, or locating may be unstable if the circuits break.

- *Failure during emergency:* Another main challenge with the sensor's need for external power is that your RTLS solution is vulnerable at times when disasters happen and the power is down. The RTLS may be down at a time when it's needed most.

✔ **Sensor calibration:** In some RTLS technologies — for example, in an Ultra Wideband-based RTLS in which alignment with respect to amplitude and phase-in-time domain is required — you may need to do location sensor calibrations during installation, and routine calibrations may be needed as well.

Deployment tools

Without the tools that provide you complete visibility into the network and help you optimize the performance, you may want to understand how the vendors and some technologies work based on the knowledge of some specific individuals and the number of times you might have to install/reinstall and adjust the system. For example, if you're using a radio frequency (RF) RTLS, an RF spectrum analyzer can give you a detailed idea of what frequencies are in use and what can be added to the mix.

Prior to the actual deployment, the proper type of software tools can tell exactly how the system will behave once deployed. Professional deployment tools, such as site survey tools with full RTLS deployment support, are required to achieve optimal results.

Estimating Maintenance Requirements

As time passes, you need to address wear and tear (by replacing tags, for example). You also need to grow the system to continually adapt to the needs of the organization using it. The system must be able to accommodate new applications and expand to handle a growing coverage area, including additional locations and more tagged objects. Here are just a few maintenance requirements to consider:

- ✔ **Troubleshooting:** Usually incorporating troubleshooting support (to identify exceptions or faults and debug or perform root-cause analysis) is more of a vendor choice rather than the RTLS technology, but some technologies are easier to troubleshoot than others. The better the acceptance of underlying RTLS technology, the better the availability of troubleshooting tools.

- ✔ **Upgrades:** Another differentiating factor is the ability to upgrade software used in tags, location sensors, the location engine, and so on. Again, this is related less to the technology and more to the choices made by the vendor.

- ✔ **Hardware replacement:** Hardware does fail. Because in an RTLS you have a large number of tags and location sensors, if any piece fails, you need to replace it. The ease of this process needs to be discussed and addressed long before the installation occurs.

- ✔ **Battery replacement:** Most of the time, the battery choice — replaceable, rechargeable, or easily available — is a vendor choice rather than the RTLS technology, but it can be part of the RTLS and it plays a prominent part in maintenance planning.

- ✔ **Tag cleaning:** Another differentiating factor is the ability to clean tags under normal cleaning conditions. Again, this is related less to the technology and more to the vendor choices.

- ✔ **RTLS expansion:** *Expansion* refers to the ability to add more tags and/or expand the RTLS into new areas. Some RTLS are easier to expand than others. The ones that are harder to install are often harder to expand.

Knowing Your Integration Needs

A fully interoperable RTLS can seamlessly integrate with other enterprise software systems, such as material, patient, warehouse, or other operational management systems, which can be enriched with real-time location information. Here are just a few questions to consider:

- ✓ **Is there any data that you might have to duplicate in the RTLS from other software systems?** For example, for an asset-locating application, you might have to duplicate information about assets into the RTLS.

- ✓ **Does the RTLS offer an easy way to import and export from your existing and planned software systems?**

- ✓ **Does the system offer application-programming interfaces to dynamically (such as runtime) interact with your other software systems?**

- ✓ **Do you know of any other software systems that will benefit with the location information?**

End users can continue using software systems that they're already familiar with if you can integrate location information into those systems.

Measuring Return on Investment

The total cost of the RTLS solution includes software, tags, location sensors, servers for the location engine or applications, deployment costs (such as cabling and planning), and maintenance costs. The direct costs of hardware, software, and installation are easy to measure. However, it's more difficult to measure the hidden costs in maintaining the system. Here's an outline of all costs that must be considered when deploying an RTLS solution:

- ✓ **Hardware costs:** Hardware costs consist of the RTLS tags as well as (optional) location sensors and other corresponding hardware.

- ✓ **Software costs:** Software costs include the end-user application software, middleware, and any licensing costs for software that runs in tags. Many vendors base the license cost on the number of tracked items, so the cost scales linearly with the actual benefit and the financial return of the system to the organization.

- ✓ **Installation costs:** Location sensors (and tag) installation adds to the system cost. If the location sensors installation requires cabling or wiring, for example, the installation may interrupt daily routines in the positioning area. For example, in a hospital setting, the installation of additional equipment and cabling disrupts daily operations and sometimes compromises patient safety. In some cases, installing extra equipment doesn't necessarily provide 100-percent read accuracy due to the orientation of the tag, the distance between the tag and the reader, or the interference, which may prevent positive identification of the tag. Furthermore, some industries require regulatory inspections and approvals in areas with any type of system installations that involve drilling holes, moving ceiling tiles, or installing low- or high-voltage cabling.

If your organization uses location sensors, maintain an additional hardware inventory of spare infrastructure as well as mounting brackets, cabling, termination panels, and other hardware in the event the system layout needs to be altered or any part of the proprietary network fails or is rendered inoperable.

✔ **Maintenance costs:** When looking at the cost of operating an RTLS solution over several years, the ease of use, reliability, and system maintenance become critical. These include planning and labor costs for replacing a failed tag or a failed location sensor; updating tags, sensors, the location engine, the middleware, or the application software system; and additional costs that are added corresponding to the organization needs, including additional locations and more tagged objects.

Other maintenance considerations include the system's ability to accommodate a newer RTLS technology that may get invented after deployment.

To protect the RTLS investment for years to come, the RTLS must have a technology-transparent architecture that can support multiple-positioning technologies and the addition of newer technology.

Considering Value Adds

Besides the accuracy, size, battery life, coverage, deployment, environment, cost, and serviceability considerations, value adds set one RTLS apart from others. The following sections describe a few of these value adds.

Tag value adds

Besides the technology that enables the location engine to locate the tags, some RTLS tags can have features, such as a push button, a connector, an LED, a buzzer, an LCD screen, a vibrator, and sensors; others quite simply do not. These value-add features enable many more RTLS applications in the same setup.

If the tags have bi-directional communication in place, tags can be remotely activated and configured. The ability to remotely configure a tag or a group of tags provides flexibility for RTLS applications so that the system can change in response to evolving requirements.

Two-way communication also enables a tag to have audible or visual indicators, such as ringtones or LED lights, that can be used to help find an object. One

or more push buttons on an RTLS tag can drive workflow applications, such as maintenance requests, supervisor requests, and asset status, to name a few. Another capability of interest is the ability to send and receive short text messages. Text messaging enables a number of interactive RTLS applications, such as sending device status (for example, *utilized, needs maintenance* and so on), paging personnel, and sending security or workflow-related alerts with codes or other instructions.

Application bells and whistles

Well, I have no clear definition of bells and whistles. Basically, these are the non-essential, but often engaging features, that enhance the user interface by making it more attractive, convenient, and seamless. Things that you can consider are

- Web-based and mobile (PDA) interfaces for ubiquitous access to the RTLS

- Real-time graphical views of tracked objects on floor maps of the location area

- The ability to search for any tag, location sensor, or any other piece of information in your RTLS in a manner similar to an Internet search

- Automated processes for pushing event alerts and escalations to end users

Evaluating Security

I don't need to describe in great detail the well-known damage and productivity loss caused by malicious hackers today. And RTLS, due to the wireless nature, is susceptible to hacks and attacks. For example, a malicious entity may be able to disrupt the location engine's ability to compute location without being in the facility, violate the privacy of legitimate users, or track their movements without being in the facility.

You need to understand how easy it is for an attacker, hobbyist, or someone unintentionally to challenge your RTLS's

- **Availability:** Loss of RTLS availability implies that the tags are no longer locatable or that they're not locatable with the expected resolution, accuracy, or performance.

✔ **Integrity:** Loss of RTLS integrity implies that either the tags are being located incorrectly or the information sent to the tags or received from the tags can't be trusted.

✔ **Confidentiality:** Loss of RTLS confidentiality implies that private information has been made available or disclosed to unauthorized individuals, entities, or processes.

I included various security threats and countermeasures that you may want to consider before selecting the RTLS solution or technology in Chapter 15. I also include vulnerabilities of different RTLS technologies in Part IV. It's very important for you to consider all the possible threats and available counter-measures with each technology and RTLS solution so that you can make the right choice.

Don't fool yourself into thinking that only a high-end hacker with sophisticated hardware and software can attack. Because of a large number of open-source software and cheap hardware, attacks can be launched by spending less than $20.

Tracking Performance

Like measuring costs, quality, quantity, cycle time, efficiency, productivity, and so on of products, services, and processes in other parts of an organization or business, having ways available to measure the RTLS performance is crucial.

You need to have mechanisms to not only distinguish one RTLS from another but also within the same RTLS so that you know

✔ How well you're doing

✔ Whether you're meeting your goals

✔ Whether your users are satisfied

✔ Whether your processes are in statistical control

✔ Whether and/or where improvements are necessary

The ease with which you can monitor one RTLS over another is an important contributing factor in deciding which RTLS to use. If you can't measure, you can't succeed.

Requiring a Standards-Based Solution

Is the system based on industry standards or proprietary technology? If based on proprietary technology, how actively is the vendor working with other vendors to standardize it? Will you be locked into a vendor's proprietary system and pricing for the long term? These are important questions to ask.

Standards are drafted by experts in the field through open participation, which accounts for their strength. Standards promote common understanding, facilitate interoperability, and encourage greater competition. If you use standards-based technology, you'll have lot more negotiating power for price and a higher ability to manage risk in the event of vendor failure.

Ensuring Eco-Friendliness

Environmental concepts, such as hazardous waste reduction, product recycling, and green supply chains, are increasingly prominent in business. Initiatives are being undertaken by people in businesses around the world to indicate an awakening of ways to reduce their immense environmental footprint. They're trying to mitigate all that by being eco-friendly — from the buildings they construct to the supplies they use.

A different RTLS adds a different level of hazardous or nonhazardous waste. You may want to compare RTLS based on the levels of pollution they add per year. For example, an RTLS that uses nonhazardous regular alkaline batteries has less hazardous waste than the RTLS that uses lithium, mercuric, oxide, nickel-cadmium, nickel metal hydride, or silver oxide batteries. Similarly, the RTLS that necessitates battery replacement every year produces more waste than the RTLS that calls for battery replacement every four years.

Making Your Choice

Because every facility is different and RTLS use will be widespread and long term, evaluating RTLS solutions using theoretical-decision matrix models may not be enough. I always advise that you do at least one of the following:

✔ **Run a proof of concept in your environment.** Proof of concept is all about running the RTLS in a small subset of the facility to see whether it really works for you. This is about making an informed decision without getting fully invested in it.

Just running the proof of concept has a financial impact because it requires spending time and energy; however, it has the potential to demonstrate how systems differ with respect to your key objectives, deployment, maintenance, and other issues as described earlier in this chapter.

✔ **Visit existing installations.** Because there are a lot of new vendors and technologies and because it may not be feasible or practical to run proof of concept at your own facility, it may be worthwhile to visit other facilities that are running the same technology.

Part II

Implementing RTLS in Your World

The 5th Wave   By Rich Tennant

Setting up RTLS check points has helped our business.

In this part . . .

The real value of an RTLS isn't the technology, but rather the application of the technology in real-world practice. This part contains four chapters to help you do just that.

In Chapter 4, I stress upon the importance of understanding the real-world usage of your RTLS application and provide you some questions that you must answer before you even look at the alternative solutions or their trade-offs. In Chapters 5 and 6, you discover what to do as part of preparing for installation, what you need to do to ensure that the contracted level of service is provided to the users of the RTLS, and what processes you need in place so that you can react to failures after it's installed. In Chapter 7, I present various integration opportunities and what you may want to do to integrate an RTLS with your existing technology.

Take it from me, after you know how an RTLS will be used, installed, managed, and weaved in the fabric of your organization, you'll be more comfortable in selecting the right technology for an RTLS.

Chapter 4

Planning for a Successful Implementation

Whatever your goals for implementing an RTLS, whether you have immediate short-term goals or motivations for the long term, you'll find that adopting an RTLS is no different from adopting any other new technology or process. Visualizing your goals and planning for expansion, as discussed in this chapter, are critical steps toward ensuring a successful RTLS implementation.

Visualizing the Goals

Before you even look at what you need to succeed, visualize what you want to do. Create a mental image of what you want to happen as though it's happened already.

Visualizing helps you create strategies, enhance effectiveness, and achieve success. It also helps you overcome obstacles and conquer unexpected hurdles. Besides defining the purpose of your RTLS application, such as reducing time spent finding assets or improving asset security, visualize the following:

✔ Who will use the application?

✔ How and where will users employ the application?

✔ What will make users happy?

✔ Where will users go for help when they have problems?

✔ What don't you want your users to face?

✔ What are the signs that the application isn't working well?

I cover all these topics in the following sections.

Although no scientific instruments can measure the power of visualization, I have no doubt that the importance of creating and writing down your mental images of the RTLS in action is necessary to fully complete the goal for your RTLS implementation. For example, if you're planning to deploy an asset-locating application, you can create images of its usage scenarios. For instance:

✔ The locating application should be available in the handheld devices carried by your staff.

✔ Only authorized persons can locate, so there must be password protection.

✔ The locating application should be easy to use with no training needed.

✔ The locating application should have at least three buttons: to locate a specific asset, locate all assets, and locate asset in-recall.

✔ Your staff person should be able to specify range — only assets within a specified distance should be reported.

✔ The screens should have big fonts and big buttons so that the staff person can hold at arm's-length.

✔ When the application reports the asset's location, it should be reported graphically on the floorplan and so on.

A clear mental image allows you to brainstorm the application with actual potential users as well as detect new opportunities and avoid dangers.

Identifying your users

When you plan an application, first you must identify the users (or groups of users) of that application. Identifying your users allows you to plan access to the resources that users need, create the right user interfaces for these people, and develop the proper workflow models.

To identify your users or user groups, ask the following questions:

- ✔ Can you identify a user or user groups for the specific RTLS application?
- ✔ Do you know when users might access the application?
- ✔ Do you know what other applications these users use or need?
- ✔ Do you know what applications aren't needed by these users?
- ✔ Do you know which other users or user groups use information from these users?

If the answers to all these questions are "yes," you can begin to plan your application, as you know what your target user groups are and what kind of applications are required. If the answers to some questions are "sometimes" or "maybe," however, you may find it helpful to use a systematic approach to identify your user groups. A systematic approach simply involves writing down all the user groups that could possibly use this application and then finding out the detailed answers for these questions from a few users in each user group.

If you're planning a staff safety application, for example, the users will be (a) the staff members who carry RTLS call-button tags and (b) the security staff members who will respond to alerts from users' tags. Anything that doesn't apply directly to your users isn't going to be successful.

Addressing how users will use the application

The next step in visualization is defining the interface your users will use. By interface, I mean how your users will interact with or use the RTLS tags and application.

Think about all the users who will use the application and how they'll use it. If you're considering staff safety, for example, visualize the following elements of interface:

- ✔ **The form factor of the tags:** The shape, color, and size of the tags and what the call button looks like.
- ✔ **How the tags will be used:** Worn as pendants, carried in pockets, included in employee badges, and so on.
- ✔ **How the button operation will work:** What should the sequence of operations to raise and/or clear the alert be? Just pressing the button may raise an alert, for example, or pressing the button for a few seconds may be necessary to raise the alert.

✔ **The security staff user interface:** Devices such as personal digital assistants (PDAs) or desktop computers, as well as the graphical user interface of the application. Security staff members may see the location of the alert (the person who pressed the call button to request assistance), for example.

✔ **Tag density:** The maximum and average number of RTLS tags that you expect to have in an area. In the security example, the RTLS must be able to track the location for alerts pressed by the maximum number of people at the same time.

✔ **Tag speed:** The rate at which your tags could be moving in the area. In the security example, a user could be walking or running when he presses the call button; the RTLS must be able to locate the tag (once the button has been pressed) when the tag is moving at maximum speed.

You also need to visualize any other variables that are specific to your application. In staff safety, for example, one such variable is *response time:* the period from the time when a user presses the call button to the time when a member of the security staff contacts that user.

Knowing where users will work with the application

In a staff safety application, for example, you need to identify the facility (or locations within the facility) where users can press the call button, as well as the locations that security staff members monitor for alerts.

Asking what makes users happy

What goals must your application meet to satisfy users? One way to get data is to meet users in focus groups. Another way is to assume that you're in the shoes of your users and visualize the following aspects of the application:

✔ **Little or no training required:** Users shouldn't have to be trained beyond basic operation. In a security application, users shouldn't have to be taught how to press the call button, and security guards shouldn't have to be taught how to use the screens when the application starts.

✔ **No location-specific constraints:** There should be no constraints with respect to where users can or can't press the call buttons.

- ✔ **Maintenance:** The care and maintenance instructions should be clearly specified so that users can understand the impact.

- ✔ **Clear indicators:** The tags should have clear indicators of when they're working (or not working), as well as clear indicators of when they're activated. For example, a tag might blink with a green LED when activated or blink with a red LED when running on low battery.

Using the RTLS isn't the primary job of your users. An RTLS should be as nonintrusive and easy to use as possible.

Supporting your help staff

A group of users that is often ignored but that plays a crucial part in the success of your application is the help or support staff — the people who are contacted whenever anything related to the RTLS, whether it's a tag or software, doesn't work as users expect. You need to ensure that you identify your support staff and represent them in your selection and installation of an RTLS.

Knowing what you don't want

While you visualize the who, how, what, and where of your application, also list the things that you don't want with the RTLS implementation.

Here are some examples (but as I mention earlier, all requirements are specific to your needs):

- ✔ You don't want to change tag batteries every three months.

- ✔ You don't want to deal with installing cables or wires through the walls or outdoors.

- ✔ You don't want even a single result for a tag location that's inaccurate by more than 30 feet.

Understanding the Business Constraints

In addition to analyzing your goals and determining your needs for an RTLS, you need to analyze any business constraints that may affect your choice and operation of an RTLS. The following sections discuss typical business constraints that you may encounter when you implement an RTLS.

Working within budget and staffing constraints

Your RTLS design must fit the budget, which should include allocations for hardware (tags, location sensors, and other hardware for running the location engine and application software), software, software licenses, maintenance and support agreements, testing, training, and staffing needs.

To ensure the success of your project, determine who controls the network budget, such as the information technology (IT) department, security department, material managers, network managers, or other departments. Also determine how much control users and user groups have over expenditures.

Regardless of who controls the budget, one common goal for the RTLS is containing costs. You should develop a Return on Investment (ROI) analysis and explain how the deployment of an RTLS will pay for itself through reduced operational costs, improved employee productivity, improved employee retention, or increased revenue potential and market expansion.

Analyzing the abilities of in-house staff members is also a good idea. You don't want to limit your choice of an RTLS to the technologies and protocols that your users understand. This analysis will determine your needs for additional personnel, consulting, training, or outsourcing needs.

Accommodating the technical landscape

It'd be nice if you could escape discussing technological religion (technology preferences), but avoiding this topic can put your RTLS initiative at risk.

You must have discussions about protocols, standards, and vendors. Keep in mind the following crucial elements:

- ✔ Any preferences (standardization policies) on transport, routing, desktop, or other protocols.
- ✔ Any doctrines regarding open versus proprietary solutions.
- ✔ Any policies on approved vendors or platforms. In many cases, a company has already chosen technologies and products for the new network, and your design must fit into the plans.

By having brainstorming sessions around these topics, you can build a consensus that the RTLS technology in question won't be disruptive. This is almost a mandatory requirement before decision makers make a go or no-go decision.

Be sure that you know what technologies are already deployed or being planned for your facility or facilities. You don't want to pick a technology that precludes any other planned application or interrupts any existing application.

Offsetting environmental conditions

You must design your RTLS for the environment in which your RTLS technology (tags and location sensors) must operate. Following are some key questions to ask:

- ✔ **Conditions:** What are the humidity, pressure, and temperature conditions in which the tags will operate? Does the RTLS technology or tag hardware operate at optimum level in those conditions?

- ✔ **Environment:** Does the environment contain extensive metal, such as steel structures? Will the RTLS technology under consideration produce high accuracy and consistent results in those environments? Some RTLS technologies, such as Wi-Fi, suffer from high signal reflections in such environments and location accuracy may suffer.

- ✔ **Dust:** Is the environment dusty? This dictates the choice of enclosures you need for the tags. You'll need dust-proof packaging for your tags.

- ✔ **Dynamic:** Is the environment dynamic (changing frequently)? This affects your choice of technology because the RTLS technology that computes location is based on scene analysis.

- ✔ **Cleaning requirements:** Do the tags have any specific cleaning requirements? For example, solar cell-based tags will need regular cleaning because the accumulated dust decreases their ability to gather energy.

- ✔ **Cleaning restrictions:** Do the tags have any specific cleaning restrictions? For example, tags affixed on medical equipment might need to be able to withstand regular disinfectant procedures.

- ✔ **Deployment challenges:** What are the challenges involved in deploying location sensors? For example, it may not be practical to deploy cables or location sensors in large open areas.

Understanding business processes

An RTLS isn't just a technology; it's also a business process. You need to have a clear understanding of the existing workflows so that you understand where and how your application will make the process better. Supplementing current manual processes or automated systems with an RTLS is effective only if you target problem areas successfully and identify how you will measure improvements. In a staff safety application, for example, you'd identify typical response times before and after the RTLS is deployed.

In addition to process improvements, you need to understand the organization's analytics and reporting so that the state of your RTLS application is reported to the appropriate people.

Checking out your suppliers, clients, and competitors

If you know what processes and choices have been made by your suppliers, clients, and competitors with respect to an RTLS, you can start contemplating what might be beneficial to you also. Following are the questions that you want to ask:

- ✔ **Vendors:** What are your vendors doing in RTLS projects? For example, if one of your vendors is using it to track recalled items, does it make sense for you?

- ✔ **Clients:** Are your clients using an RTLS? For example, if the first thing your client does when he receives products from you is tag them, maybe you need to look into tagging as a value add or for your own benefits.

- ✔ **Competitors:** What are your competitors doing? Watch all press releases and marketing announcements to figure out the reasons behind competitor choices. Some might be applicable to you for your business.

- ✔ **Technologies:** What technologies are being used by your vendors, clients, and competitors? A simple reasoning is that they have teams of people doing the investigative work, and you want to benefit from their research.

Understanding why and how your vendors, clients, or competitors are embarking on RTLS projects will help you envision how you can use an RTLS successfully as well.

Enlisting Support

An important part of performing due diligence is knowing exactly who will benefit from the RTLS implementation beyond the primary users. What potential exists for other departments? Identify all the primary constituents and educate them on the benefits of an RTLS.

In the long term, an RTLS has enormous benefits and can affect an entire operation: manufacturing, operations, packaging, warehouse management, security, research, billing, materials management, finance, IT, and so on. Secure buy-in from upper management, and form a cross-functional team

consisting of folks from all departments, whether or not the initial RTLS application affects them. Educate the team on real RTLS performance and expectations, and leverage any data collected internally among departments.

Make sure that you have a clear definition of how success is defined by the executives, the team, and any other stakeholders. Also determine whether the definition of success will change as yearly fiscal goals change.

Engaging allies early helps you build the support you need to sustain your RTLS initiative over the long term, as well as achieve a better ROI by using the same infrastructure for many applications.

Defining the Scope of Your RTLS Implementation

By scope, I mean developing a common understanding as to what's included and what isn't included in your RTLS application. For example, if you're creating an asset-security application that will alert you when an asset leaves a well-defined perimeter, define things such as

- ✔ **Areas:** The areas (or perimeters) where asset security will be implemented.
- ✔ **Expectations near perimeter:** What's expected when the asset is seen near the perimeter? For example, automatic door locks, alerts to security, and so on.
- ✔ **Expectations outside perimeter:** What's expected when the asset has actually crossed the perimeter? For example, can a security guard still locate it after it's left the perimeter?

The key components of scope of an RTLS application are

- ✔ **Physical areas:** The areas where the application will work, potentially narrowing it by region, department, or smaller zone in the facility.
- ✔ **Assets/people:** List of assets/users that will be involved.
- ✔ **Functionality:** The features of the application.

Scope will change as you proceed, and this is to be expected. It's difficult to foresee all needed features at the start, and as the details become clearer and people figure out how to use the RTLS, more features will creep in. For these reasons, you need to have some contingency (such as additional time and money) built in your plans.

For a first project, try to minimize scope. For example, instead of implementing asset security for the whole hospital, implement it first only in the emergency department. Grow your RTLS in stages. Planning small projects will increase your chances of success and lead to wider RTLS adoption throughout the organization.

Anticipating the Future

Although you need to be realistic about what you can accomplish in the initial phases of RTLS implementation, don't shoot yourself in the foot by making choices that limit the possibilities of the RTLS infrastructure. The things you may want to consider are

- **Types of assets:** If you have assets in different shapes and form factors, you may want to think about how you'll be attaching the tags to those assets. A good strategy is to think about all high-value assets in your organization, whether or not you target them in your initial implementation.

- **People locating:** Your RTLS implementation may or may not track people, and your needs for tracking people may be different in different phases of deployment, but do think about applications such as evacuation management (making sure that all employees or visitors have left the building and assembled in predetermined evacuation areas) and visitor tracking.

- **Accuracy:** In the initial implementation, you may not need very high accuracy, but you may want to plan so that you don't need to install totally different technology throughout the facility if you later need higher accuracy for some other application.

- **Scope:** Based on your facility or needs, you may not need an RTLS that works both indoors and outdoors or in all areas in the facility, but you may want to plan in such a way that no part of the facility is left out.

Understanding the Technology

Understanding what technology will work best in your environment — as well as the custom technology, integration, installation, and maintenance requirements in your environment for your applications both today and tomorrow — is one of the most important tasks with respect to an RTLS.

Besides investing in building your staff's expertise in the technology, you can do the following things:

✔ Invite RTLS hardware, software, and integration partners to demonstrate their products' capabilities to users.

✔ Participate in RTLS events and organizations, such as RFID Journal Live, IDTechEx — Active RFID, RTLS & Sensor Networks, and so on.

✔ Visit sites and partners that have working RTLS solutions.

Find partners who know and support more than one technology so that the partners won't be hung up on a specific RTLS technology and you will get more unbiased advice.

Creating an Implementation Plan

A high-level implementation plan is the final step that sets the stage for success. The plan consists of resources, milestones and deadlines, project tasks, and dependencies.

✔ **Resources:** Resources refer to all the equipment, materials, and personnel required for the project. Selecting personnel with the right skill set for different tasks is very important for success.

✔ **Milestones and deadlines:** These are the events (or dates) that indicate significant progress or are stages of completion of major tasks. For example, completing the installation of each location sensor is a good milestone to track and completing RTLS implementation fully is the final deadline.

✔ **Project tasks:** These are all the tasks that you have to do, such as planning where to install location sensors, actually installing location sensors, and so on. Each task is usually defined with all the resources.

✔ **Dependencies:** Dependencies are very important to recognize: You can't start using the tags until you've installed location sensors. You can't install location sensors until you've made the decision on a vendor and technology. And so on. Other dependencies are things such as the nature of certain tasks — the number of personnel involved, their specific needs, and so on.

You can use scheduling and planning software, such as Microsoft Office Project; access a project management tool online; use spreadsheet software; or simply keep track manually, with pen and paper. No matter whether software is used, establish processes that alert you about any significant events, such as slips in the schedule.

Chapter 5

Preparing for Installation

In This Chapter

▶ Defining your test model

▶ Choosing technologies

▶ Selecting a vendor

▶ Conducting the prepilot test

▶ Running the pilot test

···

*I*mplementing a full RTLS involves a significant amount of money and procedural changes throughout an organization. Having a deployment plan that works is essential; you don't want to waste money or find yourself backtracking to correct critical errors that could have been prevented with a little planning.

This chapter tells you the right questions to ask and provides the framework for running a pilot test to ensure that you're up and running with your RTLS as quickly and easily as possible.

Adopting a Test Model

Before you deploy an RTLS, you must figure out one very important thing: Will it work?

No single test model answers this question, and your particular experience will vary based on many factors, including availability and skill set of your own internal resources, sense of project urgency, support of decision-makers, and project complexity.

Generally speaking, though, the following four steps can help you determine the right technology and vendor:

1. Select a list of suitable technologies.

2. Select a list of suitable vendors.

3. Conduct a prepilot test to evaluate multiple vendors and make the final selection.

4. Conduct a pilot test before full deployment.

The ultimate goal of a running a test model is making a "go or no-go" decision on implementing the full pilot.

Selecting Technologies

When you select technologies, you gain a baseline understanding of various technologies, as well as a list of the technologies that are suitable for your applications and environment (and eliminate the ones that are not).

You should think of this phase as the phase of playing around with the technology and understanding its applicability to your applications and environment. Typically, you facilitate this process by obtaining starter kits from one or more RTLS vendors and using them to test how the RTLS works in an office or lab environment.

If you're planning to locate assets indoors, for example, you eliminate standard Global Positioning System (GPS) technology in this phase because standard GPS requires a clear view of the sky to lock onto and track satellites. GPS signals travel by *line of sight,* meaning that they pass through clouds, glass, plastic, and other transparent objects but not through solid objects, such as buildings and mountains; as a result, GPS typically can't locate indoors, underground, in urban canyons, or in similar environments. More details on various technologies as well as their pros and cons are in Part III.

Because playing around with technology allows people to get some hands-on experience, it helps you develop in-house knowledge. Organically grown in-house talent can support your prepilot and pilot tests better, as well as your actual deployments. Accelerating the learning curve and growing this talent, even if you bring in seasoned consultants, will probably pay off in long term.

Interviewing Vendors

The success or failure of an RTLS initiative rests on the selection of the appropriate technology vendor. The goal of this phase is to prepare a list of vendors that provide the suitable technologies for your purposes.

In this phase, you hold brainstorming sessions within your organization as well as with different vendors to understand how their products and solutions fit your application and environment. The more rigorously you question your true needs, the better able you are to select the best products and vendors to evaluate.

When reviewing RTLS solution choices from several vendors, keep in mind these questions (independent of the technological choice):

✔ **Has the vendor's system been thoroughly tested in different environments (and especially in an environment like yours)?** Unless the technology you're looking into is a recent innovation with significant theoretical advantages, it usually is in your best interest to work with a solution that has been deployed in many different environments as well as in environments like yours. The key advantages include stability of the solution (more bugs have been fixed), mature troubleshooting capabilities (the vendor has figured out how to debug and resolve issues faster), and a better road map to the solution (the vendor is getting ideas and suggestions from a diverse pool of customers).

✔ **Has the system been tested on different types of assets (especially assets like yours) and carried in different ways by people?** A specific tag technology, tag design, or asset behavior can cause undesired results in the short term as well as the long term. A tag meant to be carried by a person, for example, may seem to be ergonomic and reasonable but may not be really comfortable to wear. Alternatively, a tag may have some radio frequency (RF) technology that interferes with the RF communications of the asset itself, or the asset may create significant vibrations that prevent the tag from performing optimally. You can prevent these problems and many other pains if the solution you're planning to explore has been tested and used by other assets or people in similar ways.

✓ **What are the costs of installation and deinstallation for a small area as well as for the full deployment?** Often, for reasons beyond your control, you may not have full understanding of installation costs in the beginning, and incremental costs may occur while the installation proceeds. By having this data for both small and large areas, you may be able to extrapolate the variation in costs for the complete installation based on the variations of cost in a smaller pilot installation. One cost that usually isn't discussed enough is the cost of deinstallation; significant deinstallation costs may hinder your expansion and change plans.

✓ **Can the solution be integrated with many enterprise applications, especially applications like yours?** If the RTLS solution under consideration has been integrated with many applications, it probably has much more mature application programming interfaces (APIs) for the software, which makes integration with your application easier.

✓ **Can the technology be upgraded?** Some vendors provide tags and hardware that aren't upgradeable in the field. Having non-upgradeable tags can be an advantage because such tags usually are hackerproof, but if a critical bug occurs in your environment, you may be left with no option but to remove and reattach all tags. For this reason, you may prefer upgradable tags. Similarly, if the software in location sensors isn't upgradeable, you may end up uninstalling and reinstalling all the hardware in your ceilings, walls, and so on.

✓ **Does the vendor have a broad tag portfolio?** A *broad portfolio* means that you can expand your RTLS deployment to different areas and different applications; it also signifies the vendor's maturity. For example, if the vendor has tags that can be deployed indoors only, you may not be able to use the tags in outdoor areas. Or, if the vendor has only one shape/size tags, you may not be able to use those on different types of assets.

✓ **Can quality-assurance procedures be established, and does the vendor have some reports and performance metrics to bring to the table?** Having quality-assurance metrics indicates the maturity of the solution as well as the vendor's experience.

After you select the technologies and vendors, the next logical step is conducting a prepilot test.

Testing the Waters with the Prepilot

The *prepilot* phase (often referred to as *proof of concept*) is the phase in which you invite vendors to prove that their solutions work in your environment for your application. This phase allows you to evaluate technologies and vendors, and then make the final selection.

Prepilot tests are useful for providing a feasibility study of the technology and performance data on the technology, as well as for giving you a high-level idea of your installation needs and vendors' capabilities. A prepilot test confirms not only the applicability, but also the practicality and cost-effectiveness of a technology and the RTLS solution.

During the initial prepilot phase, you should select a small group of savvy users to test the technologies, as you don't want to burden yourself with training and support.

Although you don't want to draw significant resources for the prepilot phrase, you do want to have a well-defined test scenario and performance-metrics goals. Here's the recommended approach for the prepilot test:

- **Clearly defined scope:** Do the testing in an area that's representative of the whole environment — or, if possible, in the worst environment for the technology that you're trying. The test area shouldn't be so big that testing interferes with your business or so small that you can't evaluate accuracy.

- **Realistic use scenarios:** Use scenarios that are as close to the business-case assumptions as possible. Select the tag types (attached to assets or carried by people) that are as close to the real business case as possible. If you're planning to invest in an RTLS for asset security to prevent loss and valuable equipment theft and want to receive alerts when assets move into and out of areas of interest, you need to have a similar setup and similar accuracy expectations in your prepilot testing area and you should conduct the testing around it.

  Establishing expectations up front about what must work and what is merely desirable in the real application saves headaches down the road.

- **Clearly defined test plan:** The basic objective of your test plan is to clearly define each test case (such as the steps to be executed and the expected results), define deliverables and responsible parties, and communicate to all responsible parties the test strategy employed. Work with your vendors to coordinate and set expectations on this test plan.

- **Simplicity:** You shouldn't have any training or support needs during testing.

- **Clearly defined performance metrics:** The performance metrics depend on your application, but you can't compare technologies unless you have clearly defined metrics and record data meticulously.

- **Multiple technologies:** Testing multiple technologies at the same time or closely spaced in time enables you to apply similar rules to everyone and understand the differences in various technologies and solutions very easily.

- **Input collection:** A side goal of a prepilot test is collecting input from prepilot users, even for the technology that you select as best. This information will help you achieve success in the actual pilot test.

Conducting the Pilot

In most technology rollouts, one of the difficult decisions is the implementation strategy: Do you take the big-bang approach and get it done with quickly, or do you slowly phase in new processes and technology over time? Although the appeal of the big-bang (or *sudden implementation*) strategy is that it focuses the organization for an intense and relatively shorter period of time than if the project were phased, in most new technology rollouts, the big-bang approach is less than optimal. After you narrow the selection in the prepilot test, lay the foundation for evaluating the RTLS technology in a pilot setting.

Conducting a pilot before full deployment is a low risk approach and is less likely to result in relative chaos. However, you do want to keep focus and conduct pilots in a very timely manner to maintain a sense of urgency and to keep within your budget because it will now take

- More time to address the problem for which you're investigating an RTLS
- More time to realize the benefits of an RTLS
- More time to see savings by use of an RTLS

The goal of the pilot test is to identify problems and enable optimization before wider deployment. Every environment is unique, but you can use the procedures in the following sections as guidelines for running a pilot program.

Identifying the tasks and resources

Identify the tasks and resources necessary to conduct the pilot RTLS program. Your objective is to clearly identify your needs with respect to

- **Computers and other hardware:** As needed for location engine, middleware, and application software.
- **Location sensors:** The location sensors needed for the pilot.
- **Tags:** The tags (and spares) needed for the pilot.
- **Assets/people:** Identifying the exact assets or persons to whom the tags will be attached or be carried by.
- **Required physical installation:** This implies understanding and planning of all physical installation, such as digging walls.
- **Required network changes:** This may mean setting up new networks, such as virtual wired or wireless local area networks, configuring *DHCP (dynamic host configuration protocol)* servers that allocate IP addresses for all network devices, such as location sensors, computers, and so on.

✓ **Required access controls:** This may mean getting access to your network passwords, security keys, and so on so that tags, location sensors, and so on can communicate as well as potentially give the vendor access for remote troubleshooting.

✓ **Business disruptions:** It's important to plan for the potential of business disruption during the installation process for the pilot.

✓ **Communication process:** This means identifying the means and mechanisms, such as e-mails, broadcast voice messages, frequency of e-mails, and so on for communicating stages, progress, and the pilot status.

Here are a few key things that you need to consider for planning your pilot:

✓ **Scope:** Create a pilot implementation of the RTLS solution in a limited, defined area. Controlling the scope, scale, and breadth of deployment is necessary for success. Don't limit the scope of system features to be tested during the pilot, however; you want to see the full application in action.

✓ **Users:** Identify and select user groups for your pilot program. These users should be at about the same technical level as your system users in general. If your organization is large or has groups with vastly different skill sets, you should run a pilot program for each unique environment or group. Locate groups that are representative of the users throughout your organization.

To maximize your success, make sure that volunteers have enough time in their schedules to participate fully in your pilot program.

✓ **Tag installation:** Plan for the processes of tag installation (often also dubbed as *tag birthing, attaching tag,* or *associating tag to an asset or person*). As part of these processes, you also want to define the steps for de-attaching tags, tag replacement, and changing batteries (if needed).

✓ **Support:** Plan to provide support for all issues, errors, or problems that users report. Develop a user-support plan that meets the needs of your users. As part of this plan, you need to identify the support staff and its resources.

You also may want to identify the mechanisms that your end users and administrators use to provide you feedback. You need to ask your support personnel these basic questions:

• How difficult is it to troubleshoot the source of a problem?

• How much support can the support staff provide remotely, without being next to the users who are facing problems?

• How easy is it to replace tags, location sensors, and other equipment, as well as to do any upgrades?

• Can end users report problems accurately?

• How easy to use are the troubleshooting tools that are available to end users?

✔ **Training:** To ensure success, you must provide adequate training for all participants. Develop a training plan that meets your users' needs. Use the training plan as the basis for the final deployment of your custom package. Revise the plan based on feedback from the pilot program and use the revised plan to train users for the final deployment. To develop this plan, you also need to identify the training staff and resources (as I discuss earlier in this section).

An important thing to note is that training for administrators and support users should be separate from end-user training so that the unique needs of each group are addressed properly. End users are more likely to need training in how to use and how to report an issue, whereas administration and support staff will benefit more from knowing how to troubleshoot.

✔ **Test plan and performance metrics:** You should create a test plan for your pilot program. Start by answering these basic questions:

- What performance metrics are committed and required for your business case, such as response times in your application and desired accuracy?

- What scenarios will you follow when the RTLS fails?

- What scenarios will you follow when the RTLS works perfectly?

- What are the expectations in the areas where the RTLS isn't deployed?

- How reliable are the equipment and tags under the given conditions?

- What is the optimal equipment configuration?

You need to include the expectations for the test plan in the training so that the users can provide desired feedback.

✔ **Rollout plan:** You should create the installation and go-live plan for the pilot for deploying the tags, location sensors, servers, and application software. As part of this process, you should identify all equipment needs and resources, the people who will be installing equipment, the installation schedule, and any effects on end users (whether or not they're participating in the pilot test).

You should also prepare a checklist and identify the feedback questionnaire that you want your installation team to provide during and after the rollout.

✔ **Evaluation criteria:** Evaluation criteria for the pilot test include the number of users who were dissatisfied, the number of problems reported, the number of support calls and requests, and the resolution rate for problems.

Getting ready and doing the rollout

When you get ready for the RTLS rollout, you must perform a few basic tasks:

✔ Provide end-user training.

✔ Provide administrator/support-staff training.

✔ Set up feedback mechanisms, such as a Web site or e-mail alias, so that end users, installers, and administrators can provide constant feedback to the design and testing teams.

✔ Get all equipment needed for the rollout.

✔ Notify all those people who will be affected by the rollout.

At appropriate milestones, such as *ready to begin installation, tag installation complete, pilot testing begins,* and so on, communicate with the team and users what you're doing so that you set the right expectations for interruptions and performance behavior.

Ensure that documentation of the installation process is very well maintained so that you can improve it as you find out more about the technology.

Conducting the pilot testing

During this phase, you should plan to provide support for all issues, errors, or problems that users report. After you make any corrections, be sure to retest the system thoroughly.

The key objective of this phase is to ensure that you get enough pilot data from your participants. The testing and feedback will help you identify the effectiveness of the pilot and give you confidence that the RTLS will indeed work in your environment.

Maintain a record of all issues and problems you encounter. These records will help you design solutions for problems. Only after making adjustments based on input from the pilot test can you begin the full deployment.

Chapter 6

Monitoring RTLS

· ·

In This Chapter

▶ Knowing what to monitor and why

▶ Understanding the different ways to monitor an RTLS

· ·

*W*hen you prepare to install the hardware and software components of your RTLS, one of the most important things you need to do is empower your administrators to streamline the time and effort required to monitor and administer it.

You may have the best technology and a perfect RTLS application, but if you don't have processes in place to ensure that the contracted level of service is provided to the users (whether people or applications) of the RTLS, to troubleshoot problems faced by your users (or inaccuracies in data received by your applications), and to anticipate or react to failures, you can't sail smoothly and succeed in your RTLS deployment.

This chapter shows you how to monitor the performance and availability of your RTLS. Armed with the information in this chapter, you can keep your environment running smoothly, provide the means to correct problems before outages occur, and reduce the costs of resolving problems. With monitoring, you can also identify trends to facilitate capacity planning.

Monitoring RTLS Applications

Most administrators understand the need for application monitoring. In fact, administrators typically monitor the basic health of application servers by keeping an eye on CPU use, system efficiency, memory use, and the like. Because an RTLS application has a large number of moving parts, however, monitoring it can be overwhelming.

When you monitor an RTLS application, you need to ensure that the following elements are operational and performing optimally (as well as anticipate any failures before they affect the business):

✔ **Tags:** Tag failures include incidents, such as hardware malfunctions and software defects. You also need to monitor the remaining battery life of each tag.

✔ **Location sensors:** Location-sensor failures include malfunctions of hardware, defects in software, and changes in the orientation or position of a sensor. If your location sensors run on battery power, you also must monitor each sensor's remaining battery life.

✔ **Accuracy of locating:** Your dynamic environment can affect your RTLS's accuracy of locating. Because locating algorithms depends on many environmental factors, you need to monitor changes in locating accuracy in all parts of the facility. Defects in software (tag, location-engine, middleware, or application software) also can cause a change in locating accuracy.

✔ **Middleware/application software:** You need to monitor the technical "up and running availability" of your middleware and application software. You also need to keep an eye on resources used by applications for optimal performance.

✔ **Server:** The server includes hardware, virtual machines, and the operating system. To ensure high availability and performance of configured applications and services, you need to do server-centric monitoring of the server hardware, operating system, and virtualized environment (if any). The values to be monitored include CPU use, memory use, networking errors, and file system limits.

✔ **Business metrics relevant to the RTLS application:** The factors you need to monitor include response time and latency.

✔ **User experience:** User experience is an emotional indicator, but it's also a good indicator of how effectively the RTLS is improving the efficiency and accuracy of tracking people and assets.

Knowing How to Monitor Your RTLS

You can establish procedures for monitoring your RTLS via a third-party managed service, manage all sites centrally, or manage each site individually. In any case, you should have an RTLS-application monitoring integrated with the overall business monitoring; for instance, monitoring the RTLS application shouldn't be a separate chore but rather part of the overall business monitoring. For example, businesses usually have an application that monitors all the critical servers and application notifications (sent alerts, such as SMS, e-mail, and so on) to the administrators (or as appropriate) whenever any of the servers become inaccessible. In this case, it's most convenient for an administrator if your RTLS application servers, location sensors, and so on are also monitored by the same application. This way an administrator doesn't have to use separate applications to monitor overall servers and the RTLS.

Monitoring your RTLS enables you to do the following:

- ✔ **Ensure 24/7 availability:** You can verify the status and availability of your application servers, middleware, location engine, location sensors, and so on. The RTLS delivers notifications to designated personnel when problems occur and helps you to quickly address situations that, left unattended, could lead to RTLS downtime.

- ✔ **Capacity planning:** You can assess and quantify resource requirements, such as network bandwidth, server cloud capacity, and so on for an RTLS and prepare for adding capacity when you add more tags or monitor at higher frequencies.

- ✔ **Reducing service costs and lower operational risk:** Automated monitoring reduces the impact of poor communication as well as need of additional resources in service and support.

- ✔ **Detecting root cause:** Facilitate locating performance bottlenecks and detecting the causes of problems.

You can monitor your RTLS in several ways, as discussed in the following sections.

Recognizing the benefits of proactive monitoring

One of the most embarrassing situations is being surprised to hear from users about a critical error or outage of your RTLS. The fact that an application is down should never be news to you. You need to ensure that your users can trust the RTLS 24/7 or as per the contract level agreement.

For this purpose, you can use *proactive monitoring,* which is a set of tests that does the following:

- ✔ Measure critical system data such as reachability and CPU, disk, or memory use on the server where the RTLS application is running

- ✔ Simulate real user experiences (exercise the components of the RTLS solution in the same way that an actual user would)

- ✔ Measure the application's performance metrics, such as response time

These measurements give you the ability to identify potential problems before they affect your users. If any value is outside the predefined threshold levels, administrators can proactively drill down to evaluate performance characteristics at the next level down.

One example of proactively testing your RTLS is deploying some tags at fixed locations and getting those tags' locations reported at specific intervals. If

performance such as response time or accuracy is outside the predefined threshold at one level, administrators can drill down to evaluate performance characteristics at the next level.

In addition, by providing consistent repeated testing of the RTLS application, you can conduct tests independent of actual end-user traffic. If a server goes down in the middle of the night, for example, and no end users are using the system at that time, the synthetic tests still identify the outage.

Because an RTLS can be affected adversely by the environment, you may want to set up proactive tests in a controlled environment to eliminate variables that may exist in a real end user's system and that may potentially skew the RTLS performance metrics. This way, you can always determine whether the RTLS is working or whether an environmental issue exists.

You can use RTLS proactive monitoring to integrate RTLS solution monitoring into your other enterprise monitoring solution quickly and easily.

Using reactive monitoring

Reactive monitoring allows you to respond quickly to issues when they arise and ensure optimal service levels more effectively. To effectively monitor an RTLS in this way, you must ensure that your RTLS solution provides you enough troubleshooting abilities, including the following:

- ✔ **Granular logs:** Your application should allow the user to establish a list of which events or categories of events should be logged. The advantage is that you can enable different kinds of logging to focus on debugging the problem at hand without bringing the whole system down or affecting everybody else in the system.

- ✔ **Log destinations:** Your application should allow you to configure different log messages to different destinations. This allows you to filter messages that are specific to the problem and find the root cause faster. This also prevents delays in the processing of critical situations because alerts for critical situations can be routed to fax, e-mail, SMS, and other external methods of communication.

Armed with better logging and error messages for your RTLS, you can pinpoint error conditions or other problems quickly, determine what is causing the problems, and resolve those problems.

Encouraging end users to participate

No matter how fancy the RTLS tags and technology you choose are, an RTLS application can't succeed without your end users' participation.

If your RTLS application is malfunctioning or underperforming, the results can be disastrous for end users, who often have priorities other than making your RTLS application work. When an end user experiences a slow RTLS or is unable to use the RTLS application to achieve the desired task, he or she loses faith not only in the application, but also in the value of the RTLS itself. This experience and its long-term aftereffects are exceptionally difficult to correct.

To prevent negative experiences among your end users and to enable your administrators to respond to end users' problems, you should engage your end users not only for interactive feedback, but also for process definition and improvement. To help head off problems, prepare your end users to do the following:

✔ **React to inaccuracies:** Explain to users what to expect and what not to expect, as well as how to report unexpected inaccuracies or behaviors.

✔ **Respond to failures:** Explain to users what is a failure and what is an acceptable behavior, as well as how to report such failures or behaviors.

✔ **Report opinions:** Encourage end users to report their opinions and problems.

If administrators and users continually share ideas and best practices, this communication will lead to increased value and sustained success of your RTLS application. Throughout the installation and after installation, this type of communication also provides an outline of various end-user communication activities that will be conducted along the timeline.

Establishing baselines

You can get valuable performance data on an RTLS solution you're monitoring, sometimes within hours of installing and activating the solution. But the data you get will be more valuable if you have historical performance data to provide context.

If your RTLS solution automatically creates statistical baselines for each relevant measurement by capturing actual data based on the time of day and day of the week, the application can establish variable performance thresholds based on real-world use. You may not need thresholds on every measurement, of course; you may need to be alerted only for out-of-boundary conditions on measurements that really matter to the user experience or to overall system performance.

After you establish thresholds that are relative to the baselines, you can configure your RTLS application to generate events or automated actions when those thresholds are exceeded.

Chapter 7

Integrating RTLS

· ·

· ·

These days, dozens — if not hundreds — of different types of open and proprietary applications for planning, managing, and monitoring materials, employees, processes, and customer relationships make up the operating system of a company. These applications support a variety of business functions, such as manufacturing, supply-chain management, financials, projects, human resources, and customer-relationship management.

When you add an RTLS to this mix, the expectation that your RTLS applications can use the enterprise data from these applications — and the reverse — is already set. Integrating an RTLS is all about integrating the real-time visibility (of tag location) seamlessly into business practices, as well as ensuring that the existing or new applications can use the RTLS data to trigger meaningful transactions.

In this chapter, I show you the benefits of integrating an RTLS with the applications and software systems your company already has in place as well as how you can make it happen.

Understanding Integration

Integration is the process of enabling applications to share data and work together as seamlessly as possible. It often involves providing new application programming interfaces (APIs) to existing applications. To be fully integrated, applications must follow new business logic and process rules as well.

Integration is a transformational project that involves people, processes, and technology. It's also a way of thinking — an approach to doing applications.

Unfortunately, integration is easier said than done. Whether applications are homegrown, purchased, or a combination, they're built over time in different languages, using different technologies, and run on different hardware platforms and operating systems with inconsistent user interfaces. As Figure 7-1 shows, companies have islands of business functions and data, with each island existing in its own problem domain. In many cases, the original architects of these applications never intended to share data. These applications were developed as shortsighted tactical actions to solve specific business problems, and many times, they were specified, funded, operated, and maintained on a limited budget.

Furthermore, it isn't a simple matter for an enterprise to discard its existing applications and develop new ones, or even to overhaul its established business processes, to effect a change in its business model. These kinds of changes are financially expensive to undertake and daunting in terms of human resources.

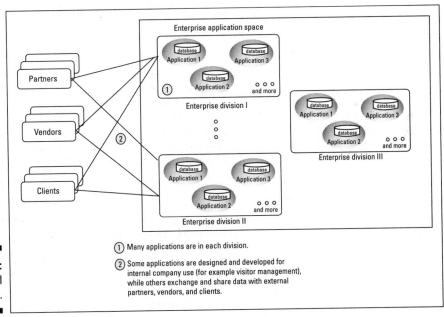

Figure 7-1:
A typical
enterprise.

When you select an RTLS solution, make sure that you're not adding a *vertical silo* — a system that's manageable and usable within its context. To monitor and manage RTLS devices such as location sensors, for example, IT personnel shouldn't have to use a separate dashboard provided by the RTLS solution; rather, they should be able to use their common dashboards and existing processes. To make an RTLS part of the fabric of the enterprise, you need to integrate the RTLS solution seamlessly into your existing enterprise middleware, databases, applications, workflow, and other systems.

Effective RTLS integration can result in many important benefits, as described in the following sections.

Streamlining inefficient procedures

Procedures that require users to switch among applications usually are more error prone and inefficient compared with procedures that involve user interaction with a single application. Integration can simplify business processes and workflows.

Suppose that you've installed an RTLS solution to locate assets, and you're also using material management software to track the maintenance history and maintenance due dates for assets. Without integration, to find assets on which maintenance is past due, you need to create a report from your material management software and then enter that data manually into the RTLS asset-locating application. With integration, you have access to maintenance due dates from the RTLS asset-locating application itself, so you can locate the assets on which maintenance is past due without going through two applications or making manual data entries.

Eliminating the cost of managing redundant data

Storing the same data in multiple databases increases hardware, software, backup needs, and administration costs; it also introduces inefficiencies and risks data corruption.

Consider a business deploying an RTLS asset-locating system. Furthermore, assume that the business is already making use of an asset management system that has information, such as manufacturer, model, serial number, purchase date, maintenance due date, and so on about all assets.

- ✓ **Problem:** Because people usually recognize assets by their model or manufacturer instead of a *serial number* (a unique number typically assigned by manufacturer), a *SKU* (*stock keeping unit,* an identification, usually alphanumeric, used to track for inventory purposes), or some other identification number, it's of limited or no value if the only way RTLS asset-locating systems refer to the assets is with their tag IDs.

- ✓ **Solution:** To solve this problem, RTLS asset-locating systems typically maintain a database that has some information about the asset, such as its manufacturer, model, serial number, maintenance due date, and so on along with the tag ID. These bits of additional information about assets are copied from the asset management system. You can note the duplication of data that can potentially result in inconsistencies, errors, and increased backup needs.

For example, imagine a human error in entering the maintenance due date, and now, the asset has two different maintenance due dates. On the other hand, if the RTLS asset-locating system was integrated (instead of maintaining a duplicate database), it actually programmatically obtained the information about assets from the asset management system, so you wouldn't have this problem.

Integration eliminates the cost of storing and managing redundant data. Overall integration also lowers ongoing system administration expenses.

Adding new business value

RTLS integration is the most cost-effective way to add new business value to the enterprise by means of innovation and new business logic and processes. Consider the following examples:

- ✓ Improve customer service by prompting the nearest customer service agent (wearing RTLS tag) when a customer summons for help.

- ✓ Ensure service quality by monitoring the attendants (wearing RTLS tags) in a senior-care facility to make sure that they are indeed making rounds.

- ✓ If an employee time-card management application is integrated with an RTLS, employees no longer need to enter data manually. Because the application is linked to the RTLS, it automatically knows when employees enter and exit the building.

✔ If a hospital's asset-disinfection management application is integrated with the RTLS, employees no longer have to sign off manually when disinfection of an asset is complete. Working with the RTLS, the asset-disinfection management application knows whether the asset went to the disinfection room after it was in a patient's room.

Increasing scalability

As your company grows, its application needs change with respect to the volume and velocity of the data that it should manage. *Scalability* refers to the ease with which an application can accommodate such changes. Many applications were designed with a specific purpose in mind and without enough emphasis on increasing functionality down the road, so their scalability is limited. Managing scalability changes across multiple applications can be a difficult task, as you may end up modifying applications one by one and piece by piece.

For example, consider a business using an asset management system and an RTLS asset-locating system for managing about 5,000 assets (similar to the example discussed in the earlier section, "Eliminating the cost of storing and managing redundant data"). If the business decides to scale the asset management system to manage over 100,000 assets and if the two applications aren't integrated, the decision makers might not see the needs of scaling the servers used by the RTLS asset-locating system at the same time. It's easier to visualize the needs and efforts for scalability of an integrated view of multiple applications. Integration lays a framework that provides an enterprise greater scalability.

Unifying device management

One of the fundamental jobs of network administrators is network monitoring and management: monitoring and managing the computers, systems, and services that comprise the network. This function allows network administrators to maintain a robust network and even to improve the network. It includes checking for slowing or failing components, as well as upgrading software and configurations as needed.

Because an RTLS solution potentially adds new hardware (such as tags and location sensors) and software (such as that used inside tags, location sensors, location engines, and middleware), a nonintegrated RTLS forces administrators to monitor several disparate dashboards. As a result, an administrator may ignore an event that's critical for location accuracy but not for network operation, or forget to apply a security patch to the RTLS. Integration enables a uniform methodology for monitoring and managing devices.

Knowing the Requirements for Integrating an RTLS

Integration usually is achieved through some form of plumbing of data transport, data transformation, and routing (see Figure 7-2). In the RTLS context, the middleware ensures data cross-pollination: use of raw RTLS events by the applications and linkage of tag and device information (such as asset IDs, personnel IDs, and location sensor IDs) from existing systems to the RTLS solution.

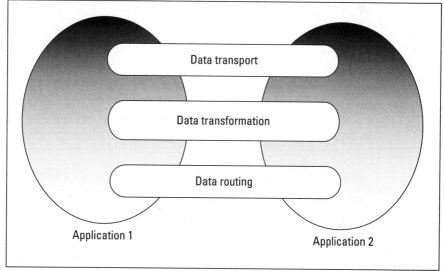

Figure 7-2: Integration plumbing.

The key components of integration plumbing are

 ✔ **Data transport:** The fundamental pipe that transfers data between software application subsystems. Data transport involves

   • Protocols (such as TCP/IP) used to move data between applications

   • Software (such as Java code) that actually transports the data

 ✔ **Data transformation:** The conversion of data from one data format (source) to another (destination). In one scenario, the source data format is the raw RTLS location events (a raw byte array, for example), and the destination data format is the data desired by an application (XML notation, for example). Data transformation involves

- *Mapping of data elements from source to destination:* This mapping establishes the linkage between the source and destination applications. RTLS tag IDs can be linked to the IDs used by existing applications, such as asset ID, employee ID, or visitor ID. This mapping is a very important part of any real-time interface because field-level differences often exist between the host and destination system, and these differences need to be mapped to achieve successful integration.

- *Translation of data from one type to another:* This translation may be very simple, such as a data element represented as an integer in one application but expected to be a string in another application. Alternatively, the translation can be quite complex, such as when translation from data element to data element requires one-to-many and many-to-one transformation rules.

- *Software code to do the transformation:* This code consists of programs/functions written in any programming language (such as C, Java, PHP, or Perl) to do that translation work.

✔ **Data routing:** The real-time communication of data (such as an RTLS location event) from one application to another.

The coordination of data from one application to another is the most difficult challenge in integration. Usually, data routing can be done in two ways:

- *Synchronously:* The application requests data periodically (also known as the *pull model*).

- *Asynchronously:* The application listens continually, and data is sent to it whenever available (also known as the *push model*).

Asynchronous architectures have proved to be the best strategy for data routing because you don't have to write APIs or function calls in one application so that they can be called by another application; in this method, the applications are loosely coupled. In asynchronous architecture, the two applications don't even have to be aware of each other's programming language. Other benefits of asynchronous architectures include less *latency* (the difference from the time when an event happens in one application to the time when the event is received by another application) and less load, as applications aren't constantly trying to communicate.

For RTLS integration, your solution should have APIs and flexible architecture. Some API specifications are described in Table 7-1.

Table 7-1	Some Specifications for RTLS Integration
Component	*Standard/Specification*
Data transport	CORBA/IIOP, FTP, HTTP, ISO/IEC 6429:1992, JDBC, JMS, OAMAS, ODBC, OSI-RPC, SMTP, SQL/CLI, Telnet, WAP, XML-Message Service
Data transformation	ANSI X.12, BizTalk, ebXML Core Components, ebXML-Message Service, HTML, JDBC, JSP, MIME, ODBC, S/MIME, SOAP, SQL, WAP–WML, WFMC Interfaces, X.12, XML Schema, XML–DTD, XQuery, XSL
Data routing	BizTalk Framework, BPEL4WS, BPML, ebXML-BPSS, ebXML-CPA, OAGIS-Integration Scenarios, UML, WFMC Reference Model and Interfaces, WSDL

Adopting RTLS solutions that support standards-based integration interfaces is the most promising way to reduce the long-term costs of integration and to facilitate a flexible infrastructure.

Performing an RTLS Integration

Like any other integration, integrating an RTLS is a complex task. Many conflicting drivers and even more possible "right" solutions exist, and unfortunately, no how-to cookbook exists for integration solutions. Also, RTLS integration is a project in itself. The implementation project is defined with its schedule based on integration flows, budget constraints, staffing needs, outsourcing needs, and other considerations.

Before you begin implementing an RTLS, you need to do an integration audit. Then you must define your integration implementation strategy, as I describe in the following sections.

Step 1: Perform an integration audit

Before you implement the integration strategy, you must perform a thorough audit of your current and future interface plan. The integration audit includes analysis of your existing and planned interfaces, along with post-audit prioritization and a report on your findings and interface strategy recommendations. It involves a comprehensive analysis of current systems, current interfaces, and current workflows.

You need to understand the applications (behaviors, interfaces, and languages) on either side and corresponding business needs to build these pipes between applications so that information can flow.

Avoid garbage in, garbage out

"Garbage in, garbage out" is *not* one of the design patterns you want to follow when designing an RTLS. You need to understand not just how to perform data transformations from one application format to another but also how to ensure that the data coming in is what you expect and that the data going out is what you plan. Otherwise, applications may ignore data that they don't understand or make wrong decisions based on incorrect interpretation.

Consider a security application that sends Short Message Service (SMS) messages for security-breach events received, and assume that this application expects the time stamp to be part of the information included in the event received. If, for some reason, the time stamp included in the event is 0, the security application may never generate the event, and the whole purpose of the application may be lost.

Step 2: Define an integration implementation strategy

An integration implementation strategy includes the road map defining the processes, workflows, and integration flows that you want to see making use of an RTLS in your organization.

Estimate the costs

One key part of analysis for defining this strategy is estimating the costs of integration, as follows:

- **Architecture cost:** The architecture cost stems from integration development, execution, and setup of the operational environment. It includes licensing, new hardware, upgraded hardware/software, and rollout costs.

- **Integration cost:** The integration cost includes development of the software for the various interfaces and communications between systems. You may be surprised if you calculate the amount of time and effort that is associated with integration, especially when the resources are highly priced (or highly unavailable).

- **Operating cost:** The operating cost includes ongoing maintenance of the integrated system.

Identify debugging and testing tools

Another key part of the integration implementation strategy is identifying debugging and testing tools for the integrated solution. It's one thing to debug a component, but when components are plugged together, they don't always fit. You need to identify debugging in the pipeline of integrated applications.

Understand integration challenges

Another aspect of the integration implementation strategy is understanding the challenges, such as the following:

- The business may have had a bad experience with IT integration in the past, or it may lack thoughtful leadership. You need to be sensitive to that situation.

- The business imperative needs to be explained properly. You must have clear objectives, know how those objectives will help the business, and how you will quantify improvements. Integration, like any other IT project, isn't an end in itself; it must serve to make the business more profitable, more competitive, or more efficient, and the value of that greater profitability, competitiveness, or efficiency must be reflected in the requirements and budget for any prospective integration project.

An RTLS is all about assets and people. Without integration, the results are isolated functionality, multiple instances of the same data, redundant manual activities, higher costs, and inefficiencies. Even if you decide to not do the integration in the initial phase of your deployment, you need to be fully aware of integration needs and costs.

Part III
Tag-A-Palooza: RTLS Technology Tour

The 5th Wave   By Rich Tennant

"It's a solid ID management and tracking system,
Ted. Over 15 years on the Kalahari and we
never lost a single lion."

In this part . . .

Today, an RTLS can be achieved using many different RTLS technologies that have different physical, operational, performance, and cost characteristics.

For anyone who is exploring technologies for an RTLS, these technologies are horses for courses. Each technology has positives and negatives, so it's crucial to understand how different technologies handle different operational scenarios — especially how they respond to adverse conditions. In this part, I give you the inside track on many technologies, if not all, to help you make informed decisions on which technology will fit the bill.

Chapter 8

Locating at Choke Points

. . . .

In This Chapter

▶ Knowing the types of choke points

▶ Using choke points in the real world

▶ Understanding the underlying technologies

. . . .

*C*hoke points are strategic structures, such as doors and other entry or exit points, that help monitor and control the flow of assets or people within a building. Security check-in lanes in airports are good examples of choke points; other examples include checkout lanes at supermarkets and exit gates in parking lots.

Choke points have also been used in military strategy to force the enemy into a narrower formation, greatly increasing military personnel's capability to control who can pass through and when.

Because choke points play critical roles in economics and security, the RTLS for locating at choke points has always received special attention from researchers and vendors.

Making Choke Points

Like any other RTLS, tags, location sensors, and a location engine are the fundamental constituents of an RTLS used for locating at choke points. A simple example of a choke point is shown in Figure 8-1. The type of location sensors and tags that are used depend upon the application using choke point. For example, if the application needs to open doors based on the tag ID, passive tags with an interrogator as a location sensor can be used, and if the application needs to lock doors based on the tag ID, active tags and a receiver as a location sensor might be required. The different types of choke points are explained in the following sections.

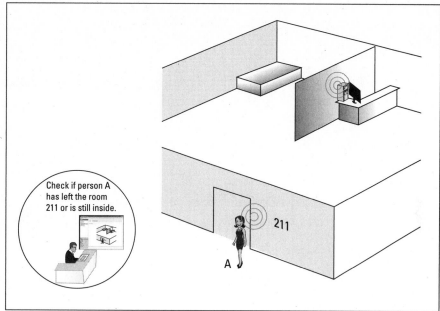

Figure 8-1:
Locating
at choke
points.

Keeping track of tags with interrogators

Interrogators (sometimes called *readers*) are short-range devices used to read passive tags by causing the tags to transmit their tag IDs and relevant info. In the interrogator model, these devices are placed at various choke points, such as interior and exterior doors and gates. For this model to work, the tags must be located within range of the interrogator, typically a few centimeters to a few meters.

You can think of an interrogator as the doorman of a choke point. Just as a doorman keeps track of people entering the building by taking down their information, the interrogator devices use radio-frequency (RF) waves or some other means to prompt tags entering its range to transmit preprogrammed tag IDs. The interrogator reads these messages and forwards them to the location engine. Then the location engine marks the location of each tag as the address of that interrogator (the choke point ID).

Interrogators are placed in such a way that the interrogation (read) range covers the choke point completely, as illustrated in Figure 8-2. If the choke point is large, multiple interrogators can be used. The tags that are attached to the assets or carried by people usually are passive.

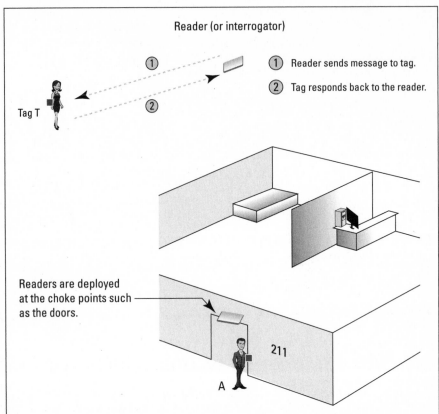

Reader (or interrogator)

① Reader sends message to tag.

② Tag responds back to the reader.

Tag T

Readers are deployed
at the choke points such
as the doors.

211

A

Figure 8-2:
Using inter-
rogators to
locate tags
at choke
points.

Locating at choke points with exciters

Exciters are transmit-only short-range devices that are used to excite the tags — that is, cause them to transmit their tag ID and relevant info to the location engine using a separate infrastructure spread throughout the facility. For this model to work, the tags must be located within range of the exciter, typically within a few centimeters to a few meters. Additionally, the exciters also include their address (the choke point ID) in their transmissions, so that when a tag gets excited and sends a message to the location engine, the address of the exciter (the choke point ID) can be included in that message. The message informs the location engine which exciter is near a tag, thereby identifying the location of the transmitting tag.

You can think of exciters as being like signposts. Just as signposts advertise information about location, exciters continuously transmit messages that include the choke point ID. When an individual or asset bearing a tag enters the exciter's range, the tag reads this choke point ID to identify the tag's location.

Exciters are placed in such a way that the range blankets the choke point. If the choke point is large, multiple exciters can be used. The tags that are attached to assets or carried by people usually have two technologies. One is the passive tag component that is excited by the exciter, and the other is a wireless network client component that communicates with the receiver infrastructure so as to communicate with the location engine, as shown in Figure 8-3.

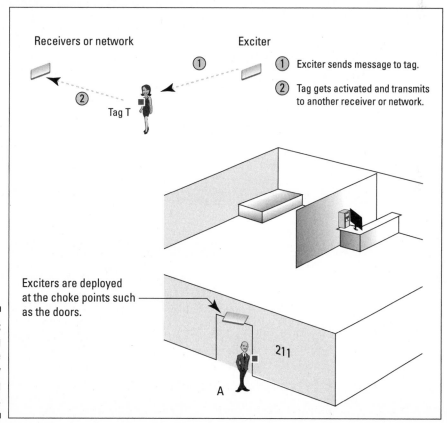

Figure 8-3:
Locating at choke points by using exciters.

Reading location with receivers

A *receiver* is a device that listens for RF signals and converts it into data packets that are available for further processing by, for example, a location engine. Locating at choke points with receivers is very similar to the way it's done with interrogators (refer to the section, "Keeping track of tags with interrogators," earlier in this chapter), except that in this case, the tags are active and continuously emit preprogrammed messages. As a result, whenever a person or asset bearing an RTLS tag enters the receiver's range, the receiver can hear the message. The receiver in turn communicates this message to the location engine that updates the location of the tag as the location of the receiver. Optionally, based on configuration parameters such as "do not allow tag ID #35 to pass through the door," the receiver can take an action such as flashing LED panels, generating SMS alert messages, locking doors, and so on.

Receivers are placed in such a way that their range covers the choke point completely. If the choke point is large, multiple receivers can be used. The tags that are attached to the assets or carried by people are active tags. See Figure 8-4.

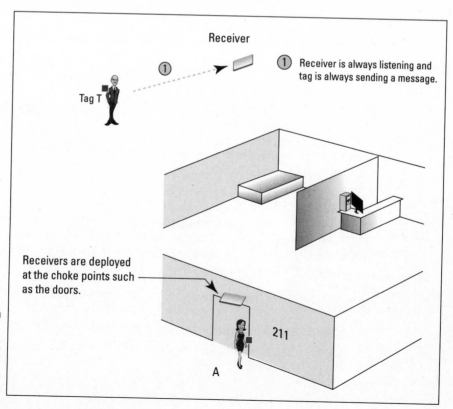

Figure 8-4:
Using
receivers to
locate tags.

Locating at choke points using precision locating systems

Precision locating systems, as described in Chapter 2, precisely pinpoint the location of tags that are attached to assets or carried by people. Because the tags are located precisely, the precision locating system can detect whether a tag is at a choke point. This model may not be suitable, however, when the accuracy of position determined by the RTLS may have a large margin of error and your choke point is small (for example, if you're using a precision locating system to determine if a person is in or out the room and your choke point is at the entrance or door of the room). Further assume that the precision locating system produces accuracy within 3–5 meters. In this scenario, if a tag is reported near the door, you may not know whether the tag is inside the room or outside the room with an accuracy range of 3 meters.

Putting Choke Points to Work

As in any other RTLS application, the use of choke points depends on the application. Choke points are used for work-in-progress applications where the tag is updated each time it passes through a choke point to indicate what manufacturing steps it has already been through, used in asset security applications, and so on.

Knowing the best uses for choke points

Because choke points make good security sense, they're good fits for applications that monitor or control the movement of assets or people into and out of a facility.

Securing door locks

A magnetic door lock at a choke point is shown in Figure 8-5. One application for this type of choke point is to prevent infant abduction from a hospital. The newborn is given a tag at birth, usually on an ankle bracelet (the mother and father or partner also have tags, and all tags have matching ID numbers), and the application activates the magnetic door lock or holds the elevator if someone tries to exit with a protected infant without authorization.

Tracking people and assets through a facility

The following list describes some applications that track people and assets by using choke points:

✔ **Inventory overflow management:** Overflow inventory in temporary facilities is difficult to track. When tags are placed on all inventory, the location of every batch and lot is visible. If the warehouse is located in a different facility, choke points can monitor when inventory left the factory and when it arrived at the warehouse. This way, you can pick and pack directly from the overflow area without accessing the main warehouse. Using choke points in overflow areas, you can also quickly tell when sufficient room is available to move overflow inventory into the main warehouse.

✔ **Evacuation management:** In any enterprise, employees or visitors can be tagged, and choke points can be established at the entry and exit points of all buildings. In the event of an emergency, when evacuation is desired, the evacuation management team can ensure that all employees and/or visitors have indeed left the building.

✔ **Food monitoring:** Various choke points can be set up at shipping, receiving, or interdepartmental doorways, and tags attached to pallets of fruits or vegetables can record where the pallets have been. This data can be used to track paths of food throughout the facility and if need be, the source of any contamination can be traced.

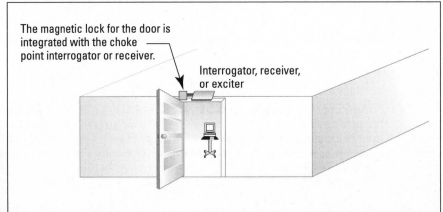

Figure 8-5:
Choke
points for
door locks.

Determining direction

Choke points can be used to determine the direction in which a person or asset is traveling. To get directionality, set up two choke-point devices (such as exciters, interrogators, or receivers) or slightly offset from one another at a choke point — for example, one on each side of a door. Based on the times when a tag was identified at the specific choke points, you can tell the direction in which the asset or person moved (see Figure 8-6).

You can also use interrogators, receivers, and so on with multiple antennas to enact two choke point behaviors.

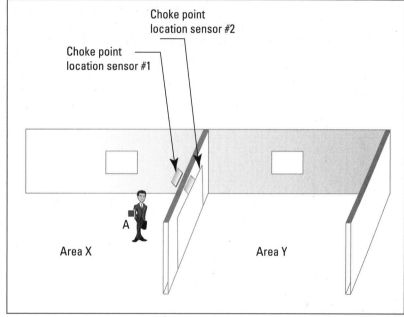

Figure 8-6:
Use two devices to identify direction through a choke point.

Managing interactions between departments

Locating at choke points can be very handy at points of interaction between workers with different disciplines or different departments, where everyone must communicate successfully. When production finishes an order and puts it in an area for completed orders, for example, the location engine can automatically notify the shipping department that the order is available at the choke point for the completed-order area. See Figure 8-7.

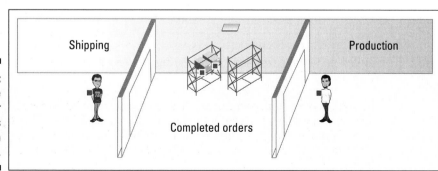

Figure 8-7:
Choke points for interactions between departments.

This system reduces the steps that production techs must take to deliver orders to shipping, and it reduces the steps that shipping workers must take to pick up the orders. Also, the system tells shipping when an order is ready, so workers don't have to make trips to production and come back empty-handed because an order wasn't ready.

Tracking the history of an asset's location

Choke points are very useful when you want to know only where an asset has been — not its exact location at all times, but what path it took and what environmental conditions it went through. Such data can play a critical role in the economics of a business. Consumers and policymakers increasingly require full and detailed information on the source, treatment, and processing of food and its components. Choke points can be used to enable food tracking in the global supply chain.

Knowing what choke points can't do

Although choke points are great for many applications, they work only if no way to get around them exists. Security guards checking the IDs of people entering or leaving a building are much less effective if somebody left a door open in the back of the building, for example.

By nature, choke points tend to get clogged. Clogs may be accidental or may occur as part of attacks. One strategy to avoid an attack or unnecessary delay at a choke point is by keeping one or more fixed reference tags at the choke point and monitoring for continuous visibility of those reference tags and raising alerts if you're missing one or more reference tags. Another strategy is by installing wireless spectrum analyzers next to the choke point and raising alerts when pre-established thresholds for noise levels are breached.

Exploring the Underlying Technologies

Several technologies enable the implementation of choke points, and in the following sections, I provide you with an overview of key technologies (except precision locating technologies) that can be used. The precision locating technologies such as Wi-Fi, ZigBee, and Active UHF are covered in Chapters 10 and 11.

I also provide you pros and cons of each technology so that you can make sure you have selected the right technology. As an example, liquids such as water absorb high-frequency radio waves, whereas metals reflect them, so if you want to track fish, you can't use choke points involving tags that use high frequencies.

Passive low frequency (passive LF)

Frequencies between 30 kHz and 300 kHz are considered to be low. Because read ranges for low frequencies typically are several inches to several feet, low-frequency-based systems make good choke-point solutions.

A typical *passive LF* (passive low frequency) choke-point system consists of passive LF tags, an antenna, and an interrogator (refer to "Keeping track of tags with interrogators," earlier in this chapter). You achieve locating at choke points by attaching passive LF tags to assets or people and placing interrogators at all critical choke points in the facility, as shown in Figure 8-8.

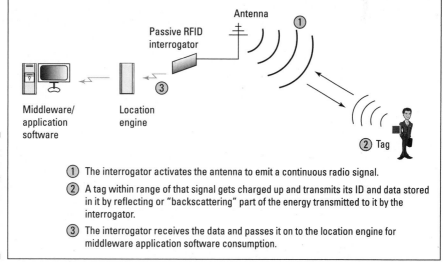

Figure 8-8:
A typical
passive LF
RTLS
system
using an
interrogator.

① The interrogator activates the antenna to emit a continuous radio signal.

② A tag within range of that signal gets charged up and transmits its ID and data stored in it by reflecting or "backscattering" part of the energy transmitted to it by the interrogator.

③ The interrogator receives the data and passes it on to the location engine for middleware application software consumption.

Passive LF RTLS systems use magnetic coupling. Such a system works this way:

1. The interrogator's antenna creates a magnetic field.

2. When a passive LF tag enters that magnetic field, an electric current is induced in the tag, powering the tag's circuitry.

3. The tag sends the interrogator a response that includes the tag ID and any other information stored in the tag.

How passive LF tags work

The magnetic field needed for coupling needs to have enough intensity to power the tag circuitry. The intensity of the magnetic field decreases with distance through a factor that can be approximated as 1 divided by the cube of the distance: $1 \div (\text{distance})^3$. This means that if you double the distance, the capability of reading the tag with the same interrogator is decreased to one eighth.

The transfer of energy from one circuit to another through a shared magnetic field is called *inductive coupling*. Usually, inductive coupling has electric and magnetic elements, but in LF systems, electric coupling is far less used. For inductively coupled tags (when used to operate passively), all the energy needed for the operation of the tags has to be provided by the interrogator.

In another model, the passive LF tags are used in conjunction with a backend wireless network client (active tag). Instead of an interrogator, this model uses an exciter and a separate receiver infrastructure to receive messages from the active tags (see Figure 8-9). (Refer to the section, "Locating at choke points with exciters," earlier in this chapter.)

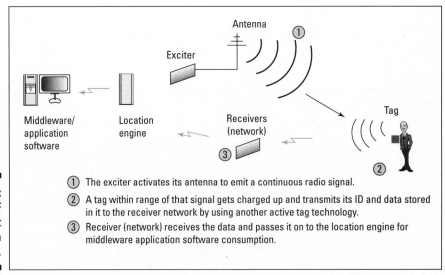

Figure 8-9: A passive LF choke point using an exciter.

① The exciter activates its antenna to emit a continuous radio signal.

② A tag within range of that signal gets charged up and transmits its ID and data stored in it to the receiver network by using another active tag technology.

③ Receiver (network) receives the data and passes it on to the location engine for middleware application software consumption.

When the LF tag gets in range of the exciter's RF field, the exciter triggers the active tag to make transmissions. Using exciters in conjunction with the receiver infrastructure provides instant information that a tagged asset or person passed through a gate, doorway, or some other choke point.

LF provides read ranges from a few centimeters to a couple of meters, depending on the sizes of the tags.

Standardization

Although LF is the oldest adopted RFID technology and has been used in several industries for many years, a little work has been done in terms of standardization in the LF arena, due mainly to the fact that most of these implementations have been in closed-loop and controlled environments.

ISO (International Organization for Standardization) has renewed its efforts to develop LF standards, however. ISO 18000-2 is the standard defining parameters for air interface communications below 135 kHz. Also, ISO 11784 and ISO 11785 have been used for some time in animal tracking. The latter two standards have several shortcomings that limit their widespread use, but they have proved to be effective where implemented. You can visit www.iso.org to find more information about these standards.

Pros and cons

Following are some of the pros or benefits of LF-based solutions:

- **Worldwide:** LF is used worldwide, with relative freedom from regulatory limitations.

- **Underwater and underground:** LF penetrates most materials, including water and body tissue, which makes it ideal for animal and fish identification. They can be used underwater and underground.

- **Environmental tolerance:** LF adapts to environments that contain metals, liquids, dirt, snow, or mud. It facilitates embedding tags in nonmetallic items (pallets, key fobs, cards, and so on) as well as it isn't affected by surrounding metals, making it ideal for use on metal items such as vehicles, equipment, tools, and containers.

- **Security:** Unlike other wireless technologies such as Wi-Fi, it has no eavesdropping risk because the eavesdropping range is the same as tag range. That means if someone is listening, that person must be close enough to be seen.

Following are some of the cons or limitations of LF-based solutions:

- **Penetration:** Doesn't penetrate or transmit around metals such as iron and steel.

✔ **Interference:** LF suffers from severely low read rates in noisy environments. The noise can be conducted on wiring, radiating short range or long range, and transient coming and going with the activation of circuits. Here is a list of sources that generate noise for LF:

- Power transformers

- Any improperly grounded electrical device

- Elevator drive motors

- Switching power supplies

- Lighting, dimmers, and lighting controls

- Electrical devices

✔ **Read rate:** LF reads at most a few tags at a time and doesn't support simultaneous reads of large numbers of tags. LF systems typically can read only 10–20 tags per second. So for choke point areas where you expect to record location of a large number of assets or people moving at very high speed, it may not a good choice.

✔ **Data transfer rate:** LF provides a relatively low data transfer rate. In general, the lower the frequency, the slower the communication.

✔ **Proprietary:** Although the LF range used is accepted worldwide, most solutions in this frequency range are proprietary, and only closed-loop applications are possible.

✔ **Large and complex tags:** Tags (and location sensors) use relatively large antennas compared with higher frequencies. Also, tags are usually more complex because of the number of turns needed in induction coils.

Passive LF tags have a very large installation base. They are used in choke point applications such as access control, asset tracking, and animal identification.

Active low frequency (active LF)

Active LF choke-point systems are different from passive LF choke-point systems because the tags used are *active* (powered by internal energy sources such as a battery). The typical range of active LF tags, like that of passive LF tags, can be from a few inches to a few feet, depending on the antenna's size, placement, and surroundings.

Because active tags don't need to extract power from a magnetic field, they don't have the distance limitations of passive tags and don't suffer from severely low read rates in noisy environments.

A typical active LF choke-point system consists of active LF tags, an antenna, and a base station (see Figure 8-10). You achieve locating at choke points by attaching active LF tags to assets or people and placing base stations at all critical choke points in the facility.

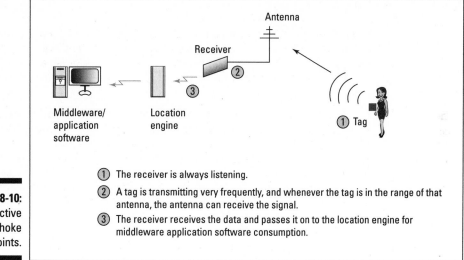

Figure 8-10:
Active
LF choke
points.

① The receiver is always listening.

② A tag is transmitting very frequently, and whenever the tag is in the range of that antenna, the antenna can receive the signal.

③ The receiver receives the data and passes it on to the location engine for middleware application software consumption.

A typical active LF system makes use of chirp tags. In this system

1. The base station is a receiver that's always listening.

2. Active LF tags *chirp* (that is, transmit their tag IDs and any other information stored in them) periodically, perhaps twice a second.

3. Whenever an active LF tag enters that magnetic field, the receiver listens to the variations in the magnetic field and thereby reads the tag response and detects the presence of the tag.

If multiple tags are present in the magnetic field of the receiver, transmissions from various tags may overlap, causing the receiver to read garbled messages. To prevent collisions and allow the receiver to read a larger number of tags, the tag ID and other information transmitted by tags must be relatively small.

Standardization

There have been no standardization efforts in active LF (specifically 30 kHz to 300 kHz) space; however, IEEE is working on 1902.1, a standard for long wavelength wireless network protocol, operating at wavelengths below 450 kHz, which will improve upon the visibility network protocol known as RuBee. RuBee typically operates in active LF (131 kHz) range and is described in the later section, "RuBee."

Pros and cons

Most of the benefits and limitations of passive LF solutions also apply to active LF solutions. However, active LF tags have a few additional pros or advantages:

- ✔ **Better read rates:** They don't suffer from severely low read rates in noisy environments.

- ✔ **Noise detection:** RF noise can block transmissions between tags and the antenna in both passive and active LF systems. In an active system that uses a receiver, however, the receiver can listen for noise and notify the system that tags may not be read because of excessive noise in the environment.

- ✔ **Less latency:** In active LF systems, because the choke point (receiver) hears the tag, the choke point can be integrated with a door controller or other general purpose IO devices, such as LED displays, robotic arms, automated gates, alarms, and so on, and the action can be done with the least possible latency. This is better in comparison to using exciters and passive LF tags where the exciters don't know if the tag is in proximity and an external message needs to be communicated to the door controller or other general-purpose IO device.

Active LF tags have a few additional cons or disadvantages as well:

- ✔ **Larger tags:** Due to the addition of battery and corresponding circuitry, battery clips, and so on, these tags are larger than passive LF tags.

- ✔ **Short battery life:** Because the power consumption of any solid state circuit is proportional to the operating speed, these tags usually exhibit limited battery life (typically between two years and a maximum of five years).

RuBee

RuBee is a two way, active wireless protocol that operates at low frequencies, always below the 450 kHz range and often in the 131 kHz range. The protocol is similar to the IEEE 802 protocols (Wi-Fi, ZigBee, and Bluetooth) in that RuBee is networked by using on-demand, peer-to-peer, active radiating transceivers. RuBee can be used in many RTLS applications such as smart entry or exit portals, animal tracking, smart shelves for high-value equipment, and so on.

A typical RuBee system includes:

- ✔ **Active LF RuBee tags:** A RuBee tag is a radiating transceiver powering its emissions using its battery and timing them using its internal crystal (clock). A RuBee tag communicates using inductive coupling, as explained in the earlier section, "Passive low frequency (passive LF)."

✔ **Base station (location sensor):** A RuBee base station is the router, and tags exchange packets (information such as their tag IDs) whenever tags come in its field. Virtually all the energy radiated by a RuBee base station is contained in the magnetic field, not the electric field. You can install base stations at choke points, and the base stations can forward information to the location engine.

RuBee protocol uses IP (Internet Protocol) addresses for tags. It's a packet-based protocol in which only one end of the communication at a time generates fields. Because the tag is a transceiver, the tags could also generate a field and establish a peer-to-peer network with other tags. This enables other tags that aren't in the field of a base station to exchange information with the base station (which forwards that information to the location engine).

RuBee has a range of 1–30 meters (3–100 feet), depending on the antenna configuration.

Standardization

IEEE is working on a standard, IEEE P1902.1 (IEEE Standard For Long Wavelength Wireless Network Protocol), with the objective to improve upon RuBee. The P1902.1 standard aims to provide a protocol that allows for the real-time searching of tags using IPv4 addresses. As part of the standard, an Ethernet-enabled router will manage the tag visibility network, allowing people to access each RuBee tag and its corresponding data over the Web.

The final specification and issued standard is expected by March 2009.

Pros and cons

Because RuBee makes use of magnetic waves and operates at low frequencies, all the pros and cons as listed in the earlier section, "Active low frequency (active LF)," apply. However, RuBee has a few additional advantages:

✔ **Peer-to-peer networking:** RuBee has the same advantage as other passive LF or active LF systems using magnetic waves — that is, RuBee works well in an active environment (with people, steel shelves, floors, cabinets, doors, and so on), as well as in situations where one or both ends of the communication are near steel or water. The key advantage of RuBee over other passive LF or active LF systems is the ability of tags to establish peer-to-peer networks, enabling a larger visibility network of thousands of tags.

✔ **Thin tags:** A RuBee tag can be nearly as thin as a credit card.

✔ **Long battery life:** Low cost 4 micrometer CMOS chips (CMOS refers to a particular style of digital circuitry design and a family of processors using a complementary metal-oxide-semiconductor), low frequency, and low speed enable extremely long battery life.

The major disadvantages of RuBee include

✔ **Speed:** RuBee operates at 1200 baud (number of distinct symbols or pulses per second), which limits its ability with respect to tag reads (typically only about 6–10 reads per second).

✔ **Small packet size:** A RuBee packet size is limited to hundreds of bytes.

If multiple tags are present in the magnetic field of the receiver, transmissions from various tags may overlap, causing the receiver to read garbled messages. To prevent collisions and allow the receiver to read a larger number of tags, the tag ID and other information transmitted by tags must be relatively small.

Passive high frequency (passive HF)

Frequencies between 3 MHz to 30 MHz are considered to be high frequency (HF), and because read ranges for HF are 1 meter or less, HF-based systems make good choke-point solutions. A typical passive HF choke-point system operates at 13.56 MHz.

The HF frequency range is accepted worldwide, so HF systems are widely used, especially in hospitals, where they don't interfere with existing equipment. HF tags were developed as low-cost, small-profile alternatives to LF RFID tags with the ability to be printed or embedded in substrates such as paper. Popular applications include library tracking and identification, patient identification, access control, laundry identification, and item-level tracking.

A typical passive HF choke-point system consists of passive HF tags, an antenna, and interrogators (readers). You achieve locating at choke points by attaching passive HF tags to assets or people and placing interrogators at all critical choke points in the facility.

Like LF systems, HF systems use inductive coupling for communication.

Although most access control systems today are based on LF, using either contactless cards or key fobs, HF is becoming the technology of choice for new access control and security systems. The additional memory allows for improved security and the integration of biometrics as part of the security features. Enhanced access control systems have the ability to validate assets as they pass through an access control system or portal. Assets with an embedded HF tag can be identified within the access control system. Documents and files can easily be identified and tracked as well.

HF has a typical range of 1 meter (3 feet), depending on the tag size and antenna configuration.

Standardization

HF (13.56 MHz) emerged in the 1990s and is the basis of numerous standards, including ISO 14443, ISO 15693, and ISO 18000-3.

Pros and cons

Following are some of the pros or benefits of HF systems:

- **Penetration:** HF signals travel well through most materials, including water and body tissue.

- **Interference:** HF isn't affected by electrical noise that may be generated by motors in an industrial environment.

- **Small and printable tags:** Because the length of the antenna is based on the length of the signal wave, the higher the frequency, the shorter the wavelength. For this reason, an antenna for a HF tag is small enough that it can be produced by printing it on a substrate in conductive ink and then affixing the chip. These tags can easily be embedded in nonmetallic items (labels, pallets, key fobs, cards, and so on).

- **Range depends upon size:** The larger the tag antenna, the greater the tag's energy capture area and the greater the communication distance from the reader. Smaller tags may be easier to package into a product, but they reduce the communication distance available.

- **High throughput:** The higher the frequency, the higher the data throughput and the faster the communication between the reader and the tags. This increase in speed allows the reader to communicate with multiple tags at the same time — a process known as *anticollision*. Compared to LF, at HF, a reader can read up to 100 tags per second.

- **Low cost:** Transponders are less expensive than LF tags.

- **Large memory:** HF tags have larger memory capacity than LF tags.

Following are some of the cons or limitations of HF systems:

- **Environmental tolerance:** HF signals are more affected by surrounding metals than LF signals are.

- **Orientation matters:** The orientation of the tags with respect to the reader antenna affects the communication range. For optimum communication range, both antennas (on the tag and the reader) should be parallel. Having the tag perpendicular to the reader antenna may significantly reduce the communication range.

- **Low data transfer rate:** HF systems have a low data-transfer rate from the tag to the reader compared to passive UHF (discussed in the following section).

Passive ultra high frequency (passive UHF)

Ultra high frequency (UHF) is the RF range from 300 MHz to 3 GHz. A typical passive UHF system operates at 915 MHz in the United States and at 868 MHz in Europe.

UHF tags boast greater read distances and superior anticollision capabilities than HF or LF (refer to the preceding sections), increasing the system's ability to identify a larger number of tags in the field at a given time.

The primary application for UHF tags is supply-chain tracking. The ability to identify large numbers of objects as they move through a facility and then through the supply chain creates an enormous opportunity for return on investment (ROI) in retail, such as reduction of wasted inventory, lost sales revenue due to out-of-stock inventory, and the elimination of the human factor required for successful bar-code data collection. Additional markets with demand for UHF RFID technology include transportation, healthcare, and aerospace.

A typical passive UHF system consists of passive UHF tags, an antenna, and an interrogator (reader). You achieve locating at choke points by attaching passive UHF tags to assets or people and placing interrogators at all critical choke points in the facility. The working principle of SAW tags is the following:

1. UHF passive tags are passive devices; that is, they don't operate under their own power (no battery). Instead, they obtain power by converting the electromagnetic energy emitted by an in-range reader to the DC power required to operate the chip.

2. The interrogator drives an antenna to send RF waves.

3. These waves propagate at the speed of light and induce a voltage in the receiving circuit in tags that are in the RF field of the interrogator.

4. The voltage is strong enough for the tags to become operational and the tag transmits information such as tag ID to the interrogator by reflecting (also called backscattering) part of the energy transmitted to it by the interrogator.

5. The interrogator receives the information about the tag.

The distance from tag to interrogator is important because passive tags don't use any traditional transmitter or battery to increase the range of this transmission. To understand backscatter, you can consider the analogy where if you aim a flashlight at a mirror, that mirror reflects the light. The brightness of the reflected light depends in part on the mirror's distance from the flashlight.

The tags are designed so that they can operate at voltages as low as 200 microwatts. Although this voltage is enough for the tags to become operational and transmit, this voltage isn't strong enough for tags to send a traditional RF transmission. Instead, passive RFID tags communicate back to the interrogator via backscatter. In the backscatter technique, the tag reflects the carrier wave and changes the load on its antenna (that is, it alters the impedance of the antenna match) to reflect back an altered signal that contains information corresponding to the data stored on the tag, such as tag ID. The interrogator's job then is to detect this change in the reflected signal. For these reasons, the passive tags usually have a short range, are lighter, have smaller form factors, and are less expensive. Several schemes are used to encode digital information from the tag to the reader (baseband signaling is the most common), but all are based on the backscatter technique.

Although a good amount of RF energy can be in an RF field, realistic applications involve human interaction, and for safety reasons, the amount of energy that a reader can emit is limited. Although these limits vary worldwide and at different frequencies, current passive RFID systems are functional typically from about 3–6 meters.

HF is a robust technology that works well for item-management applications but fails when read ranges beyond 1 meter are required. UHF targets the segment that requires longer read distances. Typical read range is 3–6 meters; however, tags with a range of 20 meters can be also created. The range depends upon the size and shape of the antenna of the RFID reader as well as the tag.

The anticollision (simultaneous reads) feature in UHF is achieved through a protocol based on bit broadcasting, as opposed to HF protocol, which operates based on the time-slot concept. This concept allows a higher number of tags to be read simultaneously in the UHF range — typically, 1,500 tags, as opposed to 50 tags in HF systems.

Standardization

One of the biggest challenges to the widespread implementation of UHF RFID is lack of globally accepted standards and regulations. In today's global economy, cross-continental trade requires goods to be identifiable universally. Currently, different frequency designations and safety regulations are in place in different regions of the world. In North America, UHF operates in the 902–928 MHz range; in Europe, UHF works in the 860–868 MHz range, whereas Japan uses 950–956 MHz.

However, some standards that have been made regarding RFID technology by standardization bodies such as ISO and EPCglobal. EPCglobal (www. epcglobalinc.org) is focused on international standards for the use of RFID as electronic product code (including passive RFID) in the identification of many items in the supply chain for companies worldwide:

- ✔ **EPC Gen2** (EPCglobal UHF Class 1 Generation 2), a standard approved for passive RFID in 2004, is the backbone of passive UHF RFID tag standards.

- ✔ **ISO/IEC 18000 - Part 6** focuses on parameters for air interface communications between 860–960 MHz.

Pros and cons

Although the UHF RFID addresses some shortcomings of the LF and HF RFID, primarily in terms of read range and read rates, it has to contend with its own limitations and challenges.

Following are some of the pros or benefits of UHF RFID:

- ✔ **Good read range:** UHF RFID provides good read distances, typically 3–6 meters.

- ✔ **Embedding:** Tags can easily be embedded in solid nonmetallic items (labels, pallets, cards, and so on).

- ✔ **High throughput:** High data throughput and a faster anticollision scheme facilitate higher read rates (typically 1,500 tags per second).

- ✔ **Low cost:** UHF transponders cost less than HF transponders due to lower memory capacity and a simpler manufacturing process.

Following are some of the cons or limitations of UHF RFID:

- ✔ **Lack of global standardization:** No global standards and regulations exist for frequency.

- ✔ **Environmental tolerance:** UHF RFID performs poorly around liquids and metals. Radio waves lose their ability to travel over distances when they encounter water and metal, which limits RFID's effectiveness in some circumstances. Liquid absorbs the radio waves, while metal can detune the RFID antenna and the reader antenna. The radio waves also bounce off metal under some circumstances, causing interference.

- ✔ **Crowded frequency band:** The 860–960 MHz range falls within the ISM (Industrial, Scientific, Medical) band, making it one of the most crowded regions of the spectrum.

- ✔ **Longer reads:** Longer read distance becomes a disadvantage in applications such as banking and access control.

Semipassive ultra high frequency (semipassive UHF)

Whereas passive Gen 2 RFID tags gather energy from the reader's signal to wake up the label's chip and provide the backscatter signal (refer to the

"Passive ultra high frequency (passive UHF)" section, earlier in this chapter), battery-assisted Gen 2 RFID tags contain an integrated power source.

This power source eliminates the need to gather energy from the reader and reach excitation, which is the primary challenge for passive tags. Typically, passive tags backscatter only 10–15 percent of the signal that reaches them. On the other hand, battery-assisted passive tags can backscatter about 90 percent of the energy that they receive.

 The battery's main purpose is to either monitor environmental conditions or to offer greater range and reliability than passive tags. You should note that the battery on the tags isn't used to generate RF energy in the semipassive tag. The battery is used solely for powering the chip and/or environmental monitoring.

Standardization

All standards that govern passive UHF RFID govern semipassive UHF also. (Refer to the preceding section, "Passive ultra high frequency (passive UHF)" for more details.)

Pros and cons

A typical semipassive UHF system is identical to a passive UHF system and all the pros and cons as listed in the section, "Passive ultra high frequency (passive UHF)," apply. However, due to the additional battery and circuitry, semipassive UHF has a few additional pros and cons.

The advantages of semipassive tags include

- **Increased reliability and read rate:** Semipassive tags have increased reliability and read rate even in the most challenging environments, such as those containing liquids and metals.

- **Sensors:** Due to its battery-based design, additional sensors such as bio-sensors, thermal sensors, and so on, can be added and included in semipassive tags.

The disadvantages of semipassive tags include

- **Batteries are required:** Semipassive tags have a battery that will need to be replaced. However, the battery has a long life, typically 10 years.

- **Higher cost:** Because of batteries and additional hardware, the semipassive tags are typically higher in cost than passive RFID tags.

- **Large size:** Due to the battery, semipassive tags are larger than passive tags.

Surface acoustic wave (SAW)

Surface acoustic waves are the radio frequency acoustic waves traveling on the surface of polished crystals. Billions of SAWs are used annually for everyday appliances, such as cell phones, pagers, TVs, and so on.

The working principle of SAW tags is the following:

1. A SAW tag uses piezoelectric crystals with reflectors at predetermined intervals to represent the tag's data, as well as input and output SAW electrode interdigital transducers (IDT) applied on its work surface.

2. The reader transmitter transmits a radio frequency pulse that is received by the tag antenna.

3. The signal from the tag antenna arrives to the tag's input IDT.

4. The IDT then produces a SAW pulse via the piezoelectric effect.

5. The surface acoustic wave travels along the SAW chip surface.

6. When the sound waves propagate along the surface of the tag, each location reflects part of the signal back. These SAW reflections from special reflector patterns produce pulses that encode the tag identification number.

7. The encoded pulse train travels back to the output IDT.

8. The reverse piezoelectric effect produces a signal on the tag antenna.

9. The encoded reply signal returns to the reader as a radio wave (and the tag ID can be read by variations in amplitude, time, phase, and/or other variables). For example, a SAW tag can be excited using a 2.45 GHz signal and can report the radio waves with 2.45 GHz as tag center frequency as well.

A SAW tag can be read up to 30 meters and located with high accuracy.

Standardization

Interestingly, not much standardization work has been done with SAW, even though SAW was invented more than a hundred years ago (Lord Rayleigh first described surface acoustic waves in an elastic body in 1885), and more than 25 years ago, the possibility of effective interaction between SAW and electrons in piezoelectrics and in piezoelectric semiconductor structures was invented using interdigital transducers (IDT).

Standardization work is in progress for SAW RFID.

Pros and cons

Because SAW tags are truly passive crystal devices, they offer many advantages over RFID technologies:

- ✔ **Environmental tolerance:** These tags can operate in harsh temperature conditions (such as less than 77 Kelvin and greater than 450 Kelvin), as well as in a hard radiation environment.

- ✔ **Better locating:** In addition to location, the direction of a tag can be determined. This enables high accuracy of order of 2 feet or so.

- ✔ **Less interference:** Because SAW tags can be read with very short spread spectrum signal pulses, many SAW readers can be deployed in the same area or work in environments where other users are operating in the same frequency band.

- ✔ **Automatic temperature measurement:** The reading process of SAW tags provides tag temperature information. Because the reflectors are precisely spaced, small variations in temperature fractionally affect the distance between reflectance pulses, which is a difference that can be measured.

Because of the nature of design, SAW tags also have some disadvantages over RFID technologies:

- ✔ **Limited tag ID set:** Because the tag ID is built into the tag, the tag ID set is very limited.

- ✔ **Tag programming:** Because the tag ID is built into the tag, the tag programming can be done only at the factory.

- ✔ **No memory:** No read/write memory is available on SAW tags.

A typical example of using SAW tags at choke points is product traceability in production processes involving high temperatures, and the use of choke-point applications for medical devices that may require sterilization.

Chapter 9

Locating at Room Level

· ·

In This Chapter

▶ Knowing how room-level locating works

▶ Locating at room level in hospitals and schools

▶ Understanding the underlying technologies

· ·

For many applications, you need to know the exact location of a person or asset; for others, your needs can be satisfied by just knowing what room a person or asset is in. This kind of location reporting — in which location is reported as an abstract idea of where something is — is called *symbolic locating*. Room-level locating is a prime example of symbolic locating; location is reported in reference to the room that a person or asset is in or near.

For many places of businesses — such as hospitals, hotels, and schools — a room identifier plays a significant role as a location indicator (for example, Operation Theater I, Waiting Room, Room #21, Exercise Room, Break Room, and so on). When you need to know where a person or asset is, a room number (or room name) is much easier to understand and find than a set of coordinates. Because of this, room-level (or subroom-level) locating has always received special attention from researchers and vendors.

Understanding the Methods of Locating at Room Level

You can use an RTLS to determine the room a person or asset is located in via a process called *locating at room level*. You can locate at room level in the following ways:

✔ **Using choke points:** You can achieve room-level locating by creating choke points at the entrance and exit of every room. Using choke points at every entrance and exit, however, requires installing location sensors at every door in the building, which can be costly and time-consuming. Also, this model can't provide subroom-level locating.

✔ **Using precision-locating methods:** You can use any of the methods that are capable of pinpointing exact position (such as the ones described in Chapters 10 and 11) to derive the room information. For example, if you know the floor plan of a building and you know the exact 3D coordinates where the person is located, you can infer the room number. The accuracy of this method depends upon the margin of error in position as computed by the precision locating method. Suppose that you can locate people inside a building within 5 meters of their actual locations. If a person is located fewer than 5 meters from a wall, the RTLS may indicate that he or she is in the wrong room or on the wrong side of the wall.

✔ **Using room-locating technologies:** You can use technologies that have characteristics that are unique to each room. For example, infrared can't penetrate walls, so if you're using an infrared tag, the infrared transmissions can be received only by the infrared receiver in that room. By knowing which infrared receiver saw the tag, the room where the tag is can be identified.

Applying Room-Level Locating

The following sections show how room-level locating can be applied in healthcare facilities and school systems to increase efficiency and ensure safety.

Increasing healthcare efficiency

Emergency department crowding is one of the major issues in hospitals and health systems that attributes to rising numbers of hours spent in waiting rooms, diversion, or bypass. One of the increases to operational efficiency can be achieved by understanding patient flow. The *patient flow* begins when a patient is first diagnosed with a particular medical condition, or when the patient first enters or is admitted to a healthcare facility; it ends when the medical condition has run its course, or when the patient leaves or is discharged from the healthcare facility. Between these two points, the patient requires a variety of healthcare resources and spends times in various rooms or areas, such as the waiting room, nurse aid station, exam room, and operating room. A hospital can improve efficiency by monitoring patient flow (from room to room) in many ways.

The following list describes a few efficiency improvements that can be achieved by adding an RTLS application to monitor (and take actions based on) patient flow:

✔ **Emergency department throughput:** This RTLS application reduces wait times by identifying potential bottlenecks before they occur. It also accommodates the high-volume, unscheduled arrival nature of the emergency department, as well as patient movement (handoffs) to other departments, such as Radiology. A tag is carried by each patient and time spent by patients in each room is monitored by an analytic application. By monitoring the time patients spend in various rooms, the hospital administration can figure out whether they need to add more staff or equipment at different stages in the emergency department.

✔ **Perioperative services throughput:** *Perioperative* generally refers to the phases of surgery that commonly include ward admission, anesthesia, surgery, and recovery. By attaching tags to patients, staff, and equipment, this RTLS application improves operating-room management and quality of patient care by reducing preoperative delays, providing fast room turnover, allowing more procedures to be performed every day, and so on.

✔ **Bed management:** This RTLS application provides automated bed (or room) management to add efficiencies in discharge and admitting, enabling better management of in-patient flows by balancing demands from the emergency department, operating rooms, and admitting physicians. In this application, patients carry tags and in one scenario alerts are automatically sent to maintenance for cleaning whenever a patient leaves a room.

Staying safe at school

Rooms are also focal points in schools; at any time, teachers and students usually are present in rooms. With increasing violence in schools, room-level locating enables quick response by security to the right room to ensure the safety of staff members and students.

The following list provides examples of RTLS applications that can help ensure the safety of students and teachers in a school:

✔ **Teacher alert:** In this RTLS application, each teacher carries a call button tag and it enables teachers to request assistance in the room by pressing the call-button tag.

✔ **Student check:** In this RTLS application, all students carry tags and presence of students in a specific room can be identified in case violence breaks out.

Knowing the Underlying Technologies

You may be able to locate at room level using choke point technologies (as discussed on Chapter 8) or any of the precision locating technologies (as described in Chapters 10 and 11). However, the three technologies that focus primarily on room-level locating are

- ✔ Infrared
- ✔ Ultrasound
- ✔ Powerline positioning

I discuss these technologies in the following sections.

For some applications, near-room accuracy may be good enough for room-level locating. And, in some applications, you may be willing to sacrifice room-level locating because your assets or personnel aren't always located in rooms. Wheelchairs in a hospital, for example, could be in rooms or in large open areas such as parking lots and parking garages, so room-level locating of those assets may not be a practical choice.

Using infrared

Infrared (IR) radiation is electromagnetic radiation with a wavelength that's longer than that of visible light but shorter than that of microwaves and terahertz radiation, as illustrated in Figure 9-1. You can't see infrared light. Because infrared light (like visible light) doesn't go through opaque barriers and reflects off the ceiling, walls, and most other objects in a typical room or enclosure, it can be used for room-level locating. For example, with an IR transmitting tag, the IR transmissions can be received only by the IR receiver in that room and by knowing which IR receiver saw the tag, the room where the tag is can be identified.

Other uses for IR

IR is used in many other applications. Night-vision equipment uses IR — when there is insufficient visible light, the radiation is detected and turned into an image where hotter objects show up brighter. Remote temperature-sensing equipment determines the temperature of objects, such as in a sauna. IR is also used in airports to remove ice from the wings of aircraft (de-icing). IR is used in short-range communication among computer peripherals and PDAs.

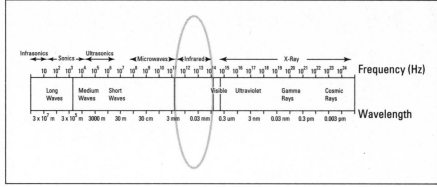

IR transmissions can be characterized as either of two main types:

- **Direct IR:** *Direct infrared* is characterized principally by the need for a line of sight (LOS) between the transmitting and receiving devices. Direct IR is point-to-point and (typically) one-to-one communication. Most consumer electronics, from camcorders to stereo equipment, include infrared remote controls. Video and audio apparatus, computers, and modern lighting installations often operate on infrared remote controls as well.

- **Diffused IR:** A *diffused infrared* device floods the room with an IR signal and then uses the reflections from the ceiling, walls, floors, and other natural surfaces to maintain robust optical communication. Diffused IR allows many-to-many connections and can be unidirectional or bidirectional. Diffused IR can create communication links at distances of 10 meters (30 feet) or more, depending on the emitted optical power.

Applying infrared technology to an RTLS

In IR for an RTLS, diffused infrared is typically used because it eliminates the need for LOS (line of sight) with the room locator (infrared receivers) to receive the IR signal from tags. Diffused IR provides exceptional robustness against shadowing and behaves like radio-frequency (RF) waves within the enclosure in which they operate. People walking around don't interfere with the ongoing diffused IR communication; neither do obstructions in the room. See Figure 9-2.

The working principle of an IR system is as follows:

1. IR receivers are installed in rooms, hallways, and (usually) ceilings.

2. IR tags worn by people or attached to equipment emit IR signals containing unique tag ID codes. Typically, the IR receivers used are highly sensitive because only a small part of the transmitted signal power from the tag reaches the receiver due to signal absorption and multiple reflections, which cause signal attenuation.

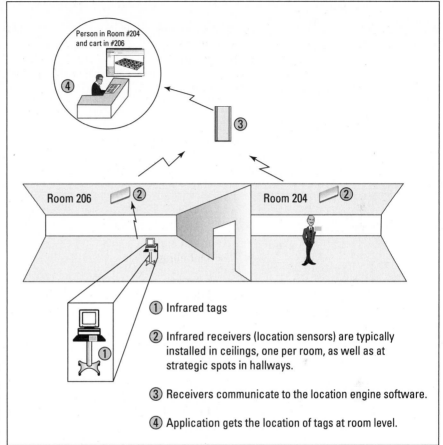

Person in Room #204 and cart in #206

④

③

② Room 206

② Room 204

① Infrared tags

② Infrared receivers (location sensors) are typically installed in ceilings, one per room, as well as at strategic spots in hallways.

Figure 9-2:
How
infrared
works.

③ Receivers communicate to the location engine software.

④ Application gets the location of tags at room level.

3. The IR receivers convert and forward the received tag transmissions (information such as tag ID) to the location engine using a wireless or wired local area network (LAN), thereby giving the application the location of that asset or person.

If multiple receivers receive the signal from a tag, the location engine can use basic triangulation (based on signal strength) to compute the location of the tag. Because the infrared beam doesn't go very far, the margin of error is low.

Because IR can't penetrate walls and ceilings and it reflects off surfaces, IR enables accurate room-level locating. Appropriate placement of IR receivers also provides subroom-level accuracy and can be focused to 12-inch accuracy.

Standardization

Although there are many standards and specifications for the short-range exchange of data over IR (for uses such as personal area networks, PANs), little work is done for standardization of using IR for RTLS purposes and many implementations are closed-loop — that is, tags and receivers must be purchased from the same vendor.

Pros and cons

Following are some of the benefits or pros of IR-based RTLS solutions:

- **Penetration:** Because infrared signals don't penetrate walls and other architectural elements, a tag's signal is received only within one room.

- **Low cost:** Infrared enables you to develop cost-effective RTLS solutions.

- **No regulations:** Infrared operates on a virtually unlimited, unregulated, optical bandwidth, so no need exists for country or state frequency allocations or any special labeling.

- **Safety:** Unlike the high RF carrier frequencies like those used by cellular and cordless phones, which have suspected health hazards, infrared is a fairly safe technology. Infrared uses the optical spectrum medium to achieve its communication and doesn't penetrate body tissues.

- **High data rates:** Infrared can achieve very high data rates that allow you to use a large number of tags or store a large amount of data in the tags.

- **Security:** Infrared light doesn't go through walls, so it's nearly impossible for an unauthorized person outside the physical controlled area of the facility to eavesdrop on the system. This situation also makes denial-of-service attacks difficult to stage.

- **No frequency coordination:** IR doesn't penetrate walls and therefore doesn't interfere with other devices in adjoining rooms. Also, it isn't susceptible to RF interference. No frequency coordination is required, as with RF systems. Receivers in adjacent rooms could be using the same IR characteristics.

Following are some of the limitations or cons of infrared-based RTLS solutions:

- **Limited range:** Infrared has an effective range of several meters, which limits cell sizes to small or medium-size rooms. In a large area, the system needs multiple infrared location sensors (at least one in each room and every few meters in hallways) to pick up the signals from tags, which could make the system expensive to install.

- **No standardization:** Due to the lack of standards or lack of conformance with standards, most infrared solutions are proprietary, meaning that the tags and receivers must be purchased from the same vendor.

✔ **Interference:** Following are some circumstances that affect the ability of the infrared sensors to accurately locate tags:

- Interference from other IR sources can come from remote controls or audio systems (these broadcast an IR signal continuously).

- Interference can also be caused by light sources such as the ballasts of fluorescent lighting fixtures. One way to limit interference is to use higher IR carrier frequencies, into the MHz region. Generally, infrared remote controls use a 32–40 kHz modulated square wave for communication.

  Direct sunlight also affects the ability of an infrared sensor to locate tags. For this reason, an infrared solution works best indoors, in areas that don't have high-frequency lights or get bright, direct sunlight.

- The infrared windows on the transmitter and receiver have to remain open to the wireless air medium. If an object covers the infrared windows, the link breaks. For tags worn by people, opacity may become an issue if the tags get obstructed by clothes or hair.

- Dark colors such as gray, blue, and black shorten the link's range. Such colors absorb infrared, so the signal can't bounce efficiently from the transmitter to the receiver.

- Plasma TVs emit strong RF and infrared signals. Unfortunately, most plasma TVs don't employ filters that filter out the emitted infrared signal, which may affect the infrared link quality.

Understanding ultrasound

Ultrasound is sound at a frequency greater than the upper limit of human hearing. Although this limit varies from person to person, it is approximately 20 kHz (20,000 Hz) in healthy young adults; thus, 20 kHz serves as a useful lower limit in describing ultrasound. Figure 9-3 shows where ultrasound fits in the spectrum of wavelengths.

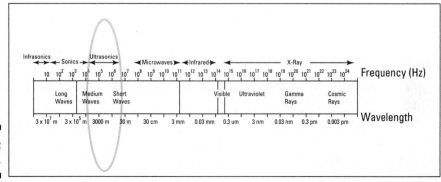

Figure 9-3: Ultrasound.

Other uses for ultrasound

Ultrasound is used in many fields, typically to penetrate a medium and measure the pattern of reflection (often called a *reflection signature*) or to supply focused energy. The reflection signature can reveal details about the inner structure of the medium. The best-known application of this technique is its use in sonography to produce pictures of fetuses in the womb.

Another common use of ultrasound is in range finding; this use is also called SONAR (sound navigation and ranging). By measuring the difference in time between the pulse being transmitted and the echo being received, it is possible to determine how far away an object is. This method is capable of easily and rapidly measuring the layout of rooms.

Because ultrasound waves can't penetrate walls, they can be used for room-level locating. For example, if there is an ultrasound transmitting tag, the ultrasounds can be heard only by the ultrasound receiver in that room. By knowing which ultrasound receiver saw the tag, the room where the tag is can be identified.

Applying ultrasound to an RTLS

The working principle of an ultrasound-based RTLS is as follows:

1. Receivers (microphones) are placed in rooms where the tracking is to be done.

2. Tags attached to the objects or people to be tracked transmit their unique identification signals using ultrasound waves. As in any other RTLS system, the tag signals are transmitted at user-defined time intervals or conditions such as detection of motion or on-demand, such when a user presses the call button on a tag.

3. When ultrasound signals, which have short wavelengths, are emitted, the walls and doors confine the signals to that room.

4. The receivers receive the transmitted ultrasound signals from a tag and transmit the signals in digital format to the location engine by using the existing wired or wireless LAN, giving the application the location of that asset or person, as illustrated in Figure 9-4.

 Ultrasound waves provide accurate room-level locating. By fine-tuning the receivers for direction, you can define subroom-level location zones, and by installing multiple receivers, you can obtain inch-level accuracy in three dimensions.

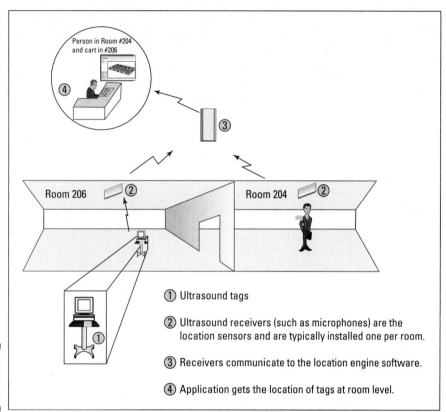

Person in Room #204
and cart in #206

④

Room 206 ②   Room 204 ②

③

① Ultrasound tags

② Ultrasound receivers (such as microphones) are the location sensors and are typically installed one per room.

③ Receivers communicate to the location engine software.

④ Application gets the location of tags at room level.

Figure 9-4:
Ultrasound
in action.

Standardization

Little work has been done for standardization of using ultrasound for RTLS purposes as of this writing. Many implementations are closed-loop — that is, the tags and receivers must be purchased from the same vendor.

Pros and cons

Following are some of the benefits or pros of ultrasound for an RTLS:

✔ **Penetration:** Ultrasound signals don't penetrate walls; therefore, an ultrasound RTLS guarantees virtually failproof room-level location accuracy. In addition, you can create subroom-level location zones, and by installing multiple receivers, you can obtain inch-level accuracy in three dimensions.

✔ **Low cost:** Because the receiver is just a microphone, ultrasound enables you to develop a cost-effective RTLS solution.

✔ **Security:** The security of an ultrasound system is very high, making it virtually impossible to eavesdrop on the communications link from outside the premises where the system is installed.

✔ **No frequency coordination:** Ultrasound waves don't penetrate walls or interfere with other devices in adjoining rooms, including sensitive equipment that might be disturbed by electromagnetic waves. Also, ultrasound waves are mechanical waves, and therefore, immune to interference. No frequency coordination is required, as with RF systems. Receivers in adjacent rooms could be using the same ultrasound characteristics.

✔ **No line of sight:** Ultrasound doesn't require LOS between the tag and the receiver, making it possible to track objects that are hidden or located in drawers or filing cabinets.

Following are some of the limitations or cons of ultrasound for an RTLS:

✔ **No global regulations:** Several countries have regulations on acceptable values of ultrasound in air, most of which rely on a statement of the International Radiation Protection Association (IRPA) from 1984. The values apply to continuous exposure of workers during an eight-hour working day. The statement recommends that people shouldn't experience sound pressure level higher than 75 decibels (dB) at 20 kHz or 110 dB at ultrasonic frequencies above 25 kHz without taking precautions. You should check sound-pressure levels with your ultrasound solution vendor.

✔ **Possible safety concerns:** As currently used in the medical environment, ultrasound poses no known risks to patients. The ear is considered to be the most ultrasound-sensitive organ, but sound-pressure levels below 120 dB at ultrasonic frequencies haven't been demonstrated to cause hearing losses. However, exposure to airborne ultrasound levels greater than approximately 155 dB results in acute harmful effects. The knowledge regarding unwanted physical effect of airborne ultrasound is quite sparse. However, symptoms such as nausea, tinnitus, headaches, and fatigue have been reported from industrial workers as a result of ultrasound stress. Another effect of ultrasound is called the *hypersonic effect,* the phenomenon reported in some scientific studies which demonstrate that although humans can't consciously hear sounds at frequencies above approximately 20 kHz, the presence or absence of those frequencies has a measurable effect on their psychological reactions.

✔ **Limited range:** Ultrasound has an effective range of several meters, which limits cell sizes (size of the area where a tag can be actually heard by an ultrasound receiver) to small or medium-size rooms. In a large area, the system needs multiple ultrasound location sensors (at least one in each room and every few meters in hallways) to pick up the signals from tags, which could make the system expensive to install.

✔ **Multipath:** Ultrasonic reception suffers from severe multipath effects caused by reflections from walls and other objects, and these are orders of magnitude longer in time than RF multipath because of the relatively long propagation time for sound waves in air.

✔ **Interference:** Quite a few ultrasound sources exist in different environments that can be sources of interference, such as ultrasound-assisted dental-cleaning equipment. Also, because you can't hear ultrasound, you may not be able to hear all sounds that are causing interference to your ultrasound RTLS, so you need to have processes in place to monitor the interference.

✔ **No standardization:** Due to the lack of standards or the lack of conformance with standards, most ultrasound solutions are proprietary, meaning that the tags and receivers must be purchased from the same vendor.

Understanding powerline positioning

Powerline communication is a system for carrying data on a conductor; this system is also used for electric-power transmission. All powerline communications systems operate by impressing a modulated carrier signal on the wiring system. Because the characteristics (such as amplitude of the signal) of carrier signals on the wires can be detected by using signal detectors and the pattern depends upon the distance of the signal detector from the source (introducing carrier signal on the wire), powerline communication can be used for an RTLS. Using powerline communication for positioning is attractive because it eliminates the need to install dedicated network cabling.

Applying powerline positioning to an RTLS

The RTLS system works based on a popular technique used to locate wires in a wall or underground. An electrician or utility worker who uses this technique connects an exposed end of the wire to a tone generator that generates a frequency range of 10–500 kHz and locates the hidden wire by using a handheld inductive tone detector.

In the powerline RTLS system, the working principle can be described as follows:

1. No location sensors are used. Instead, a few tone generators are connected to electrical outlets in the peripheral areas of the facility.

2. All tone generators continually emit their signals over the powerline.

3. The signals from these tone generators emanate from those outlets to the rest of the facility. The power of the tone is set so that even if it *attenuates* (weakens), it still reaches the opposite end (or as far as possible) of the facility.

4. Because the electrical wiring has no fixed pattern in the facilities, each area in the facility has unique signal characteristics (such as amplitude, phase, and so on). These characteristics (often called a *signature* or *fingerprint*) are recorded beforehand as part of a site survey (or scene analysis, as described in Chapter 2) and made available to the location engine. Because wiring in a building doesn't change frequently, this fingerprint remains stable for a long time.

5. Tags equipped with specially tuned tone detectors sense these signals in the building and relay that information to the location engine over a wireless network (such as Wi-Fi). The detected signal levels provide a distinctive fingerprint resulting from the density of electrical wiring at the given location. See Figure 9-5.

6. The location engine compares the signal levels received from the tag against the prerecorded database and returns the location based on the best match.

Because powerline positioning is based on fingerprints, the location accuracy depends upon the number of fingerprints and the algorithms to find the best match.

Standardization

As of this writing, little work has been done for standardization of using powerline positioning for RTLS purposes and many implementations are closed-loop — that is, tags and receivers must be purchased from the same vendor.

Powerline communication

Different types of powerline communication use different frequency bands, depending on the signal transmission characteristics of the wiring. Because the power wiring system was originally intended for transmission of AC power, the power wire circuits have a limited ability to carry higher frequencies. This ability to carry higher frequencies is a limiting factor for powerline communications.

Data rates over a powerline communication system vary widely. Low-frequency (about 100–200 kHz) carriers on high-voltage transmission lines may achieve a net data rate of a few hundred bits per second, but these circuits may be miles long.

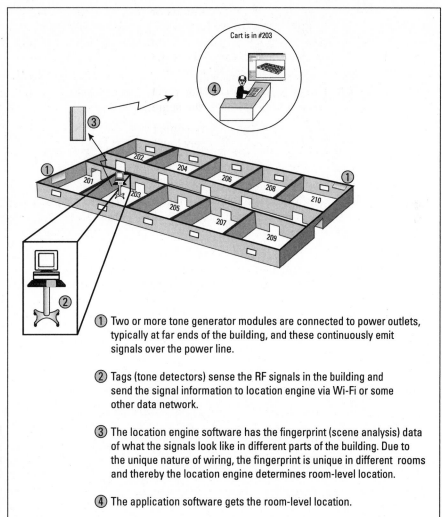

① Two or more tone generator modules are connected to power outlets, typically at far ends of the building, and these continuously emit signals over the power line.

② Tags (tone detectors) sense the RF signals in the building and send the signal information to location engine via Wi-Fi or some other data network.

③ The location engine software has the fingerprint (scene analysis) data of what the signals look like in different parts of the building. Due to the unique nature of wiring, the fingerprint is unique in different rooms and thereby the location engine determines room-level location.

④ The application software gets the room-level location.

Figure 9-5:
How powerline positioning works.

Pros and cons

Following are some of the benefits or pros of using powerline positioning for an RTLS:

- ✔ **Minimal infrastructure:** Systems based on powerline communication need very limited infrastructure additions and no cabling.

- ✔ **Low cost:** The infrastructure consists of only a few inexpensive tone generators.

Following are some of the limitations or cons of using powerline positioning for an RTLS:

✔ **Need for fingerprinting:** Because the system is based on fingerprinting, extensive fingerprinting needs to be done as part of the setup process.

✔ **RF-based:** Because powerline positioning uses low-frequency RF (radio frequency) signals, it has the same concerns as for any LF RTLS system.

✔ **Interference:** Following are some things that affect the ability of the location engine to accurately locate tags:

- Interference from nearby systems can cause signal degradation, and as a result, the fingerprint sent by the tag may not be accurate.

- Devices such as relays, transistors, and rectifiers create noise in their respective systems, increasing the likelihood of signal degradation.

- Transformers and AC-DC converters attenuate the input frequency signal almost completely. Bypass devices are necessary to enable the signal to be passed on to the receiving node. A bypass device may consist of three stages: a filter in series with a protection stage and coupler, placed in parallel with the passive device.

Chapter 10

Precision Locating with Wide Area Coverage

*P*recision locating means being able to identify the exact location of an asset or a person. You can precisely locate people or assets in many ways. You can use absolute coordinates such as latitude, longitude, and altitude (for example, a person is located at N27° 10' 0 N, E78° 2' 60 when visiting the Taj Mahal); or you can use relative coordinates, such as the distance in three dimensions from a fixed reference point (for example, the security guard is standing at 10 feet south of the main entrance of the building). You also can use relative coordinates such as distance in three dimensions from a mobile reference point. You can measure that my car is located 45 feet to the left of me, for example.

Currently, many different RTLS technologies can be used to get precise location information. These technologies can be grouped into two categories based on their *area of service* (also called as *area of coverage*). The technologies that support *wide area coverage* cover a broad area, so the tags can be located in areas that cross metropolitan, regional, or national boundaries. GPS covers the world, for example. The technologies that support *local area coverage* cover localized areas, so the tags can be located within the smaller areas such as a room, building, campus, or specific metropolitan area.

In this chapter, I discuss precision locating technologies that support wide area coverage.

Understanding Wide Area Coverage

The concept of locating a tag in a wide area is often desirable because you can locate assets or people even if they move from one facility to another, for example. Wide area coverage also implies locating tags in outdoor, urban, and even indoor environments, but typically the technologies that support wide area coverage need some assistance (or require use of precision locating technologies that support local area coverage) to provide high accuracy in local areas such as within a room, building, campus, or specific metropolitan area. The technologies that support local area coverage are discussed in Chapter 11.

Irrespective of the fact that a precision locating system supports wide area coverage or local area coverage, the precision locating systems report location in terms of distance as the accuracy level and a percentage as the precision. Some systems that use Global Positioning System (GPS) technology can locate tag positions within 3 meters for approximately 95 percent of measurements. In this example, the distance of 3 meters denotes the accuracy level, or grain size, of the position information GPS can provide; the percentage, 95, denotes precision, or how often you can expect to get that accuracy.

This grain size of locating is often used to determine how well a locating system will work for a particular application. GPS tags may suffice for species biologists who are concerned with the position of a migrating whale pod to a precision of 1 square mile, for example. A personal location system for home or office applications, however, may need enough accuracy to answer the query "Which room was I in around noon?" but not "Where, to the nearest cubic inch, was my left thumb at 12:00:46 p.m.?"

Comparing accuracy level and precision

When you're evaluating different RTLS technologies for your application, you must compare the accuracy level and precision they can provide while operating in a similar environment. To understand the accuracy level and precision effectively, you should assess the error distribution each system incurs when locating objects and the most common worst-case ways that tags will be used in your application (for example, the accuracy level and precision value in scenarios where tags could be completely submersed in water). Another aspect you want to compare is the number of location sensors (that is, how densely packed location sensors must be) to achieve the desired accuracy level and precision.

Table 10-1 describes the accuracy requirements of specific RTLS applications.

Table 10-1	Desired Accuracy of RTLS Applications
Application	*Desired Accuracy*
Automated handling	0.5 cm
Route guidance for blind	1 cm
In-building survey	1 cm
Tool positioning	1 cm
In-building robot guidance	8 cm
Formation flying	10 cm
Exhibit commentary (such as in a museum)	1 m
Pedestrian route guidance	1 m
Vehicle route guidance	1 m
Precision landing	1 m
Firefighter locating	1 m to 3 m
Train, airplane, or bus information	30 m
Social networking	3 m to 100 m

If locating systems were accurate with 100 percent probability, life would be good indeed. The practical implications (such as the ability to install an unlimited number of location sensors), however, prohibit you from installing a locating system that gives maximum accuracy with 100 percent probability.

Exploring the Underlying Technologies

Although locating in wide areas has always received special attention from researchers, the commercial availability of GPS to civilians and increased terrorism have spurred a flurry of activity in this field. The desire for a number of location identification applications that enable locating in wide area coverage has been growing dramatically.

Currently, many different RTLS technologies besides GPS (such as cellular) can be used to locate in wide areas. Because each of these technologies is designed for a specific application or a group of applications, these technologies differ in their capacity to locate:

✔ Some technologies work well outdoors; others are tailor-made for the in-building environment.

✔ Location accuracy of some technologies is in the order of 1 centimeter and for others it is 200 meters.

✔ Some technologies work well for locating underground; others work like magic for locating in water.

✔ Some technologies are multipurpose; the sole purpose of others is locating. Cellular, for example, can also be used for data, video, voice, and so on, but GPS provides only location information.

Table 10-2 lists the technologies discussed in this chapter, along with typical level of positioning accuracy using those technologies.

Table 10-2	Typical Accuracy of Locating Technologies
RTLS Technology	*Typical Accuracy of Location*
GPS	A few cm to a few m
Cellular	50 m to 200 m
WiMAX	50 m to 200 m
TV Signal	5 m to 50 m

Technological evolution is a process where the emergence of the first technology results in newer species of the same technology or newer technologies. What this means is that your vendor may offer you a slight variation, combination, or completely new technology that isn't described in this chapter. You should understand how your vendor makes use of the specific technology because that can affect the capacity of the RTLS to locate.

The following sections identify the strengths and limitations of technologies that support wide area coverage. Chapter 11 covers RTLS technologies that support locating within the local area.

Using satellite navigation systems

A satellite navigation system — also called a Global Navigation Satellite System (GNSS) — can locate people or assets with global coverage. Satellite navigation systems consist of satellites that send radio signals from space and tags that are electronic receivers. In simple terms, the tags (receivers) locate four or more of these satellites, figure out the distance to each satellite, and use this information to deduce their own location (longitude, latitude, and altitude) by using *multilateration*. Multilateration is very similar

to trilateration (discussed in Chapter 2). In multilateration, three location sensors are used for computing location and additional location sensors are used to verify and/or adjust the accuracy of result.

As of 2007, the United States' NAVSTAR GPS system is the only fully operational satellite navigation system. Due to its popularity, people sometimes refer to any satellite navigation system as GPS. Other countries, however, are developing satellite navigation systems of their own.

GPS receivers are available as standalone devices that enable personal navigation, tags that can be attached to assets, and wearable items (such as watches); these receivers are also embedded in personal digital assistant (PDA) devices, phones, and computers.

Following are a few positioning applications that make use of GPS:

- ✔ **Navigation:** Navigation applications provide personal navigation information for drivers, hikers, mariners, pilots of medical evacuation helicopters, and so on. For example tags (GPS receivers) are connected to small computers (that can display and/or vocalize) and provide street navigation for the traveler.

- ✔ **Construction fleet management:** In these applications, tags are attached to construction vehicles, equipment, machinery, and optionally, carried by people. These applications potentially reduce unauthorized vehicle use (by knowing the location of the vehicle and person and generating alerts whenever the vehicle is in an unauthorized area or is moving without any authorized person nearby), improve vehicle use by eliminating unnecessary idle time (by fast locating of equipment), eliminating dispatcher inaccuracies (by exactly knowing the location of vehicles), and so on.

- ✔ **Yard management:** These applications can improve yard-operation efficiency. For example, locating the tagged containers faster can help reduce charges customers have to pay for excess time the containers or cargo is in care of the yard authority (called *demurrage charges*) or customer (called *detention charges*); it can help in preventing spoilage or loss of trailer contents, as well as eliminate manual yard checks (locate automatically). It can also enable monitoring the performance of the equipment, the facility, and crew members.

- ✔ **Tracking of Alzheimer's patients:** Wandering is one of the most frightening symptoms of advancing dementia, and the Alzheimer's Association estimates that nearly 60 percent of patients will develop this symptom. In this application, an Alzheimer's patient wears a tracking GPS device (a GPS receiver with a back-end network radio such as cellular embedded in a watch, for example). In the scenario when the patient wanders off, the rescue teams can communicate with the tracking GPS device (using the back-end network radio) and get information about the absolute coordinates of the patient.

How satellite navigation systems work

NAVSTAR GPS is the only fully functional satellite navigation system as of this writing. In this section, I explain the working principle of GPS (see Figure 10-1).

1. A constellation of satellites (24 or more operational, plus 3 active spares) orbits the Earth every 12 hours about 20,000 kilometers above the Earth's surface. The constellation of these satellites is such that a GPS user can lock onto 5–11 satellites from any point on the Earth.

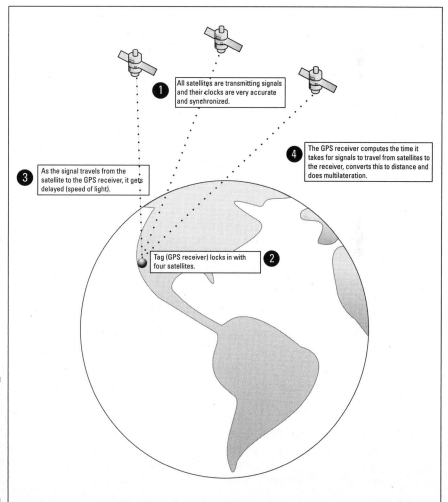

Figure 10-1:
GPS uses satellites, tracking stations, and GPS receivers.

2. These satellites transmit a code (called a *ranging signal*) on two frequencies in the microwave part of the radio spectrum — 1575.42 MHz (L1 frequency) and 1227.60 MHz (L2 frequency). (Only L1 is available for civilian use. The U.S. military can use L1 or L2.)

3. The tags (GPS receivers) listen for the signals from the satellites. The GPS receivers are time-synchronized with the satellites and run the same code.

4. The signal takes some time to reach the GPS receiver from the satellite.

5. Because the GPS receivers are time-synchronized with the satellite clock, a GPS receiver can compute the time it takes for the signal to travel from the satellite to the GPS receiver. The distance between the satellite and a GPS receiver is equal to the time it takes for the radio signal to travel from the satellite to the GPS receiver, multiplied by the speed of the radio signal (the speed of light).

6. By using multilateration with knowledge of distance from four or more satellites, the GPS receiver computes its own position.

Considering GPS safety and security

The GPS signal is weak. Measured at the surface of the Earth, GPS signal strength is about −160 dBw (1×10^{-16} watts), which is roughly equivalent to viewing a 25-watt light bulb from a distance of 10,000 miles. With the present state of scientific knowledge, it is considered to be safe (that is, nonhazardous) for people irrespective of continuous or intermittent exposure.

But because the GPS signal modulation and frequency spectrum are known to the public, GPS is vulnerable to the following types of attacks:

✔ **Blocking:** Because the signal is weak, someone can easily block it by destroying or shielding the GPS receiver's antenna.

✔ **Jamming:** Because the GPS signal modulation and frequency spectrum are known to the public, jammers can concentrate all the power within the GPS spectrum to corrupt the signals. Consequently, jammers of modest power could effectively jam the GPS signal over substantial ranges. Jamming could be unintentional, however; simple interference with the GPS signal could cause loss of the lock that GPS receivers need.

✔ **Spoofing:** Blocking and jamming aren't the greatest security risks, because a GPS receiver is fully aware when it isn't receiving the GPS signals it needs to determine position and time. A more pernicious attack, known as spoofing, involves feeding the GPS receiver fake GPS signals so that the receiver believes that it's located somewhere else in space and time. Spoofing is more sinister than jamming because it is surreptitious. In spoofing, the intended target doesn't know that the signal received from a GPS unit is wrong.

In addition to satellites and GPS receivers, a control segment that consists of *tracking stations* is located around the world to maintain GPS. These stations monitor and measure signals from the satellites to compute precise orbital data (called *ephemeris*) and satellite clock corrections for each satellite. The master control station uploads ephemeris and clock data to the satellites. Then the satellites send subsets of the orbital ephemeris data to GPS receivers over radio signals so that the receivers can compute accurate locations.

For most GPS-based RTLS solutions, another back-end network technology — wireless radio, such as Wi-Fi, WiMAX, or cellular — is typically added to the tags. This technology is used to forward the location of the receiver to the location engine (or application) on demand or periodically.

GPS doesn't work indoors, but it's a clear and effective technology for outdoor locating with positioning accuracy from a centimeter to a few meters.

Pros and cons

GPS offers an excellent solution for accurate positioning nearly anywhere on Earth, 24 hours a day, and under any weather conditions.

Here are some of the advantages of GPS:

- ✓ **Low recurring cost:** Users aren't charged for using GPS signals from the satellites.
- ✓ **High accuracy:** The positioning accuracy of GPS can vary from less than a centimeter to a few meters.
- ✓ **Privacy:** Because the tags compute their position, the information about location of a particular tag is available only to that tag.
- ✓ **Highly scalable:** Because the tags (GPS receivers) compute their own locations, no scalability issue applies for applications.

Nevertheless, despite the wide use of GPS, it has several limitations:

- ✓ **Security:** GPS is susceptible to attacks such as jamming, blocking, and spoofing.
- ✓ **Might need a back-end network:** Because the GPS receivers compute their locations, if the location of tag is needed by another application, it might need to have another networking technology (called the *back-end network*) built into the receivers. This back-end network technology sends the location (as determined by the tag) to the location engine and application. Another side effect is that if there are coverage holes (a lack of coverage in some areas) in the back-end network, the GPS tags can't be located by the application.

✔ **Line of sight requirement:** A GPS receiver has to lock onto at least four satellites to compute its location, but it can achieve lock only when it has clear lines of sight to those satellites. Because satellite signals can't penetrate solid or dense objects (such as mountains, jungles with thick branches and leaves, bridges, and buildings), a GPS receiver in such an environment may fail to locate itself.

✔ **Urban canyons:** Receivers may not be able to locate themselves in urban areas that are lined with skyscrapers (referred to as *urban canyons*), where signals can be obstructed for extended periods or continuously unavailable.

✔ **Initial delay in computing location:** GPS can require a long time to lock onto its first satellite. A traditional GPS receiver requires from 30 seconds to several minutes to lock onto and track each satellite, depending on how much information it gathered previously. This time delay can be a major issue.

✔ **Interference:** The weak GPS signal poses tremendous design challenges, especially when another wireless (back-end network) technology is also required on the same device — avoiding interference from the other wireless signal becomes a design issue.

✔ **High battery use:** GPS receivers are relatively power-intensive, making them unsuitable for long-term battery-based use.

GPS locating accuracy also can be affected by the following situations:

✔ **Satellite clock errors:** Errors in satellite ephemeris (the position of a satellite as predicted and set by control stations) result in receivers computing incorrect locations. (Receivers use satellite ephemeris along with time of travel of the signal to compute their own locations.) An error of this type is on the order of 2 to 3 meters.

✔ **Signal propagation error:** Naturally occurring variations in the Earth's atmosphere affect satellite signals by changing the speed at which they travel, which can cause time-to-travel errors.

✔ **Receiver quality:** Receiver clock quality and measurement noise can result in the receiver's providing inaccurate information.

Improvising GPS-based RTLS

Because conventional GPS has difficulty providing reliable positions when signal conditions are poor or when the receivers can't download information from the GPS satellites (rendering the receiver unable to function until a clear signal can be received continuously for up to one minute), several approaches have been taken to improve GPS location accuracy:

✔ **Indoor GPS:** *Indoor GPS* refers to locating in a wide area by using regular GPS and adding *pseudolites* — devices that act as virtual satellites. The pseudolites broadcast GPS-like signals to overcome limitations such as signal blockage for urban and indoor environments. Several pseudolites can be used and sub-meter accuracy can be achieved in indoor and urban canyon environments.

✔ **Differential GPS (D-GPS):** A *differential GPS* system makes use of differential corrections from ground-based or satellite-based augmentation systems to the GPS receiver. The ground-based or satellite-based augmentation systems basically consist of GPS signal monitors that listen to signals and send differential corrections to the GPS receiver. The correction message from ground-based monitors is normally transmitted in the 300 kHz frequency band and from satellite-based monitors is normally broadcasted in the L1 frequency band. By using these corrections, it minimizes the error contribution from the atmosphere propagation, satellite clock error, and ephemeris error so that sub-meter accuracy can be obtained. Note that D-GPS systems are heavily dependent on the communication between the GPS receiver and the augmentation system monitors.

✔ **Assisted GPS (A-GPS):** *Assisted GPS* refers to locating in a wide area by using regular GPS and an assistance server. A typical GPS receiver that uses A-GPS makes use of cellular, WiMAX, or some other data connection to contact the assistance server. With assistance, the GPS receiver can operate more quickly and efficiently because the tasks originally handled by the receiver are shared.

For example, in the case of A-GPS where cellular network is used to assist in locating, the working principle can be described as follows:

1. The A-GPS tag has a GPS receiver as well as a cellular radio (cell phone).

2. The assistance server can determine the location of tag roughly by knowing the location of cell tower to which the tag (GPS receiver and cell phone) is connected on the cellular network.

3. The assistance server can supply the orbital data for the GPS satellites to the tag's cell phone, enabling the tag's GPS receiver to lock to the satellites.

4. If the tag doesn't have a good signal from the GPS, it can pass on the fragmentary signals received from satellites to the assistance server that in turn can compute the location of tag.

Because A-GPS tags don't need to search for and decode signals from each available satellite, they improve initial delay in locking to the four satellites and this in turn also improves the battery life of the A-GPS tag. In addition, it reduces the amount of CPU and programming required on

the A-GPS tag because most of the work is offloaded onto the assistance server. A-GPS enables increased accuracy and availability, especially in urban and indoor environments. Because A-GPS is primarily designed for urban and indoor environments, it is difficult to give exact performance measures. However, with A-GPS the performance should be good enough to determine which street the car is traveling on even if the receiver can't lock in.

One downfall of A-GPS is the fact that cellular (or Internet) network assistance is required to calculate a position, making it useless in areas where no aiding information is available, such as areas without cellular coverage or compatible aiding servers in the roaming environments. Also, using network servers for position calculations makes A-GPS impractical for applications that require frequent position updates, such as continuous navigation.

Locating with cellular

Cellular networks, although not designed to track location, are attractive choices for location-based services because of the widespread use of mobile phones. In many countries, mobile phones enjoy high penetration rates — exceeding 100 percent in some countries because some people carry more than one mobile phone. Therefore, cellular network–based positioning is a highly suitable option with the potential of widespread deployment.

Following are a few positioning applications that make use of cellular technology:

- **Cell phone free zones:** Detecting and locating a person using a cell phone in a cell phone free zone such as military bases, courts, and so on.

- **People locating:** Share your location with your friends, family members, doctors, lawyers, colleagues, people in a social network, and so on, and see their location (if they enable sharing). Typically, you run a small application in your phone and use a service provided by your cell phone carrier.

- **Government security:** Detecting and locating phones installed by terrorists or other entities for eavesdropping, for example.

- **Public safety:** Locating a fugitive's cell phone, for example.

- **Emergency calls:** Providing the caller's location to the emergency first-response teams.

- **Location-based advertising:** Providing location-targeted banners and advertisements.

How cellular networks work

The cellular landscape is composed of different cellular phone networks: *1G* (the original analog circuit-switched first-generation systems), *2G* (second-generation systems that use digital encoding such as Global System for Mobile communications [GSM] and Code Division Multiple Access [CDMA]), *2.5G* (2G technology with higher speeds to support data transport such as General Packet Radio Service [GPRS]), and *3G* (data-centric third-generation networks such as EDGE). All the cellular phone networks use a portion of the radio frequency spectrum designated as Ultra High Frequency (UHF) for the transmission and reception of their signals.

The regulatory bodies of the country sell carriers a license for a block of UHF radio frequencies, which they can broadcast on in a specific area. Carriers further divide this block into smaller portions and have installed a vast network of radio towers, called *cell towers* or *base stations,* throughout their coverage area where each tower has one or more antenna that provides the coverage on its assigned portion of frequency for that site, called a *cell.* This is illustrated in Figure 10-2. The cell is roughly circular and is typically 10 square miles in size. The base stations use low-power transmitters and this allows cell towers in non-adjacent cells to use the same frequencies. Further, each base station is connected back to a central command center that connects all the cells together into a cellular network, which is linked to the worldwide phone network.

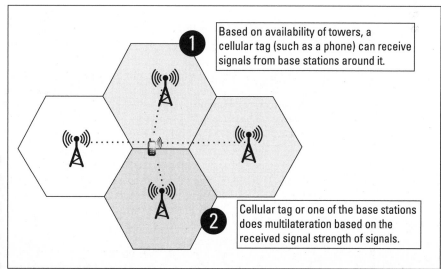

Figure 10-2:
A cellular network uses multi-lateration to locate tags.

1 Based on availability of towers, a cellular tag (such as a phone) can receive signals from base stations around it.

2 Cellular tag or one of the base stations does multilateration based on the received signal strength of signals.

Frequencies used in cellular networks

Different cellular standards use different radio frequencies. For example, IDEN uses 806–960 MHz, and GSM uses 450–496 MHz, 806–960 MHz, or 1710–2025 MHz. The cellular phones can't tune themselves automatically to the frequencies they find; they need the right hardware, not just software, to use different frequencies. For a list of the frequencies that cellular networks use, see http://en.wikipedia.org/wiki/Cellular_frequencies.

And, to allow *mobile stations* (also known as cellular phones, mobile phones, or cell phones) to connect, each base station divides its portion of frequency among a limited number of users. The total number of users a cell can support is often referred to as *capacity*. At a given location, a mobile phone can receive signals from several base stations' signals. In a GSM network, for example, a mobile phone can simultaneously receive signals from up to six base stations.

Locating of mobile stations (tags) is implemented in one of the following ways:

✔ **Network-based:** In network-based implementation, the working principle is

1. The mobile station is doing an active transmission.

2. One or several base stations make necessary measurements (such as signal strength, time of arrival, and so on) and forward this data to a central location engine (also called mobile location center).

3. Based upon the measurements, the location engine computes the location. Typically this location information is available with the carrier and it may provide the information to the end user or application for a fee. If the location information is needed at the mobile station, the location engine communicates that to the mobile station using a data (back-end) network.

✔ **Mobile-based:** In mobile-based implementation, the working principle is

1. The mobile station is neither in active call nor actively transmitting. Instead, it listens for the control channels (that are transmitted continuously) from one or several base stations.

2. The mobile station makes the measurements and computes the location.

3. The mobile station may need some assisting information, such as GPS coordinates of the base stations, to compute its location. In this model, the location information is available with the mobile phone and it may have to forward the information to another application using data (back-end) network.

✔ **Mobile-assisted:** In mobile-assisted implementation, the working principle is

1. Like mobile-based model, the mobile station is neither in active call nor actively transmitting. Instead, it listens for the control channels (that are transmitted continuously) from one or several base stations.

2. The mobile station makes the measurements but doesn't compute the location.

3. The mobile station forwards the measurements to the central location engine (also called mobile location center) in the network for further processing.

4. The computation burden is transferred to a location engine and this enables use of more sophisticated algorithms to compute the location. If the location information is needed at the mobile station, the location engine communicates that to the mobile station using a data (back-end) network.

The typical measurements (as referred to in the preceding list) used for locating in cellular networks include:

✔ **Cell identity:** This is probably the crudest way to locate using cell phones. The location of cell phone is the location of *serving base station* (the base station to which the cell phone is connected). The grain size of the location is the cell size that can be a few hundred feet (100 meters) to a couple of miles (3 kilometers), depending upon the cell size and density of cell towers.

✔ **Time advance:** In this model, the location is computed by multilateration, using round-trip delay of signal between the base station and mobile station and the location of base stations. The grain size of the location depends upon the signal frequency, typically is around 1 mile, and in the worst case, is the same as the cell size.

✔ **Received Signal Strength Indicator (RSSI):** In this model, location is computed by multilateration, using the RSSI of the signal between the mobile station and base station and the location of the base stations. The grain size of the location depends upon the distance, physical elements (such as buildings), dynamic elements in between, and geographical layout, and typically is around 1 mile, and in a worst case, is the same as the cell size.

✔ **Uplink Time Difference Of Arrival (UL-TDOA):** In this method, the location of the mobile station during a call is accomplished by forcing the phone to request a handover to several neighboring base stations.

On request for handover, the mobile station sends access bursts at full power, and the difference in time of arrival of (bursts) signal from the mobile station to base stations as measured by different base stations is used for locating. The grain size of the location is 50–200 meters.

✔ **Enhanced Observed Time Difference (E-OTD):** In this model, additional reference mobile stations (fixed at well-known locations) are used. The location is computed by a central location engine that makes use of the time difference of arrival of signals between mobile station and base stations, as well as time difference of arrival of signals between the reference mobile stations and the base stations. The additional reference mobile stations enhance the accuracy of the location engine because it can use some corrections based on real data. The grain size of the location is 50–200 meters.

Because cell network localization is less precise than other methods, this technique isn't suitable for navigation applications. Many other applications don't require a high degree of precision, so a rough position estimate can be valuable. Knowing your own rough location may be of value when you're traveling in an unknown area, for example, and knowing another person's rough location may be useful when you're deciding whether to meet up with that person, considering the distance and required travel time to the other person's location.

The location accuracy achieved in cellular networks depends upon the cell size and cell tower density. It is typically suitable for applications where accuracy better than 50–200 meters isn't crucial. Because of the large grain size, it's also difficult to locate the floor level in a building.

Pros and cons

Cellular network locating has the following advantages:

✔ **Widely available:** Cellular networks reach over 80 percent of the world's population today.

✔ **Works indoors and urban areas:** Unlike GPS, it can provide location information in urban areas, including indoors.

✔ **No additional tags for people:** Because a cell phone is the tag, no additional tags need to be carried by people. Tags are required only for assets.

✔ **Quick computation:** Unlike GPS that needs to lock in to satellites, there is no initial delay with a cellular network.

This technique also has several limitations:

- ✔ **Low accuracy:** This technique isn't well suited for situations in which precision better than 50–200 meters is crucial. Tracking accuracy also depends heavily on the size of the cell site and density of cells.

- ✔ **Dependence on carriers:** Some models of locating are dependent upon cellular network operators, not all of which may provide this service. It isn't available outside cellular networks, which implies you can't locate cellular tags in an area where the cellular network isn't available.

- ✔ **Long-term infrastructure cost:** It involves huge long-term infrastructure costs, because more cells are needed to improve the accuracy of the location information. However, there are obvious additional benefits of infrastructure deployment.

- ✔ **Recurring cost:** It may also involve the cost of requesting location information through cellular networks because many times, the computed location information is available only with the carriers.

- ✔ **High battery use:** Cellular connections are power-hungry; as a result, it's difficult to use it for frequent location accuracy (say, locating every second) for long periods with small batteries.

Improvising a cellular RTLS

In an effort to improve cellular location accuracy indoors, cellular receivers can be deployed indoors. These receivers — essentially, spectrum analyzers that detect radio-frequency (RF) activity — are independent from cellular service providers. They don't decode data packets or listen into the calls and are entirely nonintrusive. They simply collect data and send it to a centralized location engine that calculates locations based on received signal strength or the time differences in the arrivals of signals from three or more such receivers. By using these receivers, the location engine can locate a cellular device with accuracy on the order of 3–5 meters indoors. See Figure 10-3.

Because cell phones can be located indoors and outside, these receivers potentially make cell phones the only technology that can be seamlessly located both indoors and outdoors with high accuracy. You can also use these receivers to locate indoors, irrespective of carrier or phone.

Another way to improve cellular location accuracy indoors is by making use of femtocells. *Femtocells* are miniaturized cellular base stations that connect to the carrier's network via broadband (such as DSL or cable). A femtocell allows service providers to extend service coverage indoors, especially where access would otherwise be unavailable or limited. Not only do the femtocells offer improvements to coverage and capacity, but they also provide another fixed point of reference for the purposes of multilateration to compute location of cellular phones. Because femtocells have smaller coverage areas, they enable higher location accuracy.

Safety and security in the cellular network

Cellular devices use electromagnetic radiation in the microwave range, and some people believe that this radiation may be harmful to human health, possibly causing brain cancer, neurological effects, and reproductive effects. Also of interest to researchers is whether cellular radiation contributes to fatigue, sleep disturbances, dizziness, loss of mental attention, reaction times and memory retentiveness, headaches, malaise, tachycardia (heart palpitations), and disturbances of the digestive system.

To protect the population living around base stations and users of mobile handsets, governments and regulatory bodies have adopted safety standards that limit the maximum power output from a mobile phone (GSM handsets can use peak power of 2 watts, and CDMA2000 handsets can use peak power of 1 watt, for example) or base station (base station emissions for 1800 MHz are limited to 900 microwatts per square centimeter). Some standards also impose further limits on continuous exposure levels. Many national and international standards exist, and devices do operate under those safety limits, but the controversial question is whether current safety standards are adequate enough to protect the public's long-term health.

Cell networks weren't built with security in mind. In fact, to satisfy the basic need to allow people around the world to talk to one another through different systems, cell networks were designed with open standards, which impose many security risks:

✔ **Jamming:** Because the cellular signal modulation and frequency spectrum are known to the public, jammers can concentrate all the power within this spectrum to jam the wireless channel and therefore deny access to any legitimate devices in the network. Jamming could be unintentional,

however; interference with the signal could prevent cellular devices from connecting.

✔ **Spoofing:** A malicious attacker can act as a legitimate base station, thereby denying mobile station access of cellular network.

✔ **Denial of service (DOS) attacks:** DOS attacks are potent attacks that can bring down an entire network infrastructure. In a DOS attack, excessive data (such as excessive text messages) is sent to the network, more than the network can handle, resulting in cellular devices being unable to access network resources. Launching a large-scale DOS attack from a single host may be difficult, so a distributed denial of service (DDOS) attack may be carried out, with several hosts being used to launch the attack.

✔ **Eavesdropping and man-in-the-middle attacks:** The encryption algorithms can be cracked by motivated attackers. Such an attacker can intercept sensitive communication by eavesdropping, or sit between a cell phone and base station and intercept messages between them — a situation known as a man-in-the-middle attack, in which the attacker can intercept messages in both directions and change the content without the users ever knowing. Man-in-the-middle attacks are discussed in Chapter 15.

✔ **Session hijacking:** A malicious attacker can highjack an established session and act as a legitimate base station, thereby denying the mobile station access to the cellular network.

You should consider these security risks as well as countermeasures as part of your process of determining suitable technology for your RTLS application. See Chapter 15 for more about RTLS security.

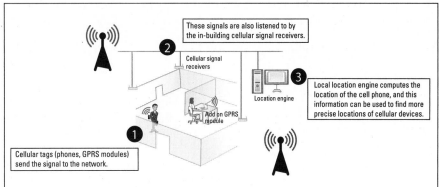

Figure 10-3:
Using a cellular RTLS indoors and outdoors.

Working with WiMAX

WiMAX (Worldwide Interoperability for Microwave Access) is a wireless digital communications system, also known as IEEE 802.16, that is intended for wireless metropolitan area networks. It can blanket a 30-mile radius with broadband access, potentially erasing the suburban and rural blackout areas that currently have no broadband Internet access. Like cellular networks, WiMAX can be used as an RTLS technology supporting wide area coverage.

WiMAX technology

WiMAX isn't a technology; rather, it's a certification or stamp of approval given to equipment that meets certain conformity and interoperability tests for the IEEE 802.16 family of standards. Wi-Fi (wireless fidelity) is a similar certification — in this case, for equipment based on IEEE standards from the 802.11 working group for wireless local area networks. WiMAX is very different from Wi-Fi in the way it works, however. WiMAX can provide broadband wireless access for up to 30 miles (50 kilometers) for fixed stations and for 3–10 miles (5–15 kilometers) for mobile stations. By contrast, Wi-Fi is limited in most cases to 100–300 feet (30–100 meters). WiMAX easily achieves Wi-Fi–like data rates that are easily supported, but the issue of interference is lessened.

WiMAX operates on both licensed and non-licensed frequencies, providing a regulated environment and viable economic model for wireless carriers. For non-line of sight service, WiMAX uses a lower frequency range: 2 GHz to 11 GHz (similar to Wi-Fi). This type of access is limited to a 4–6-mile radius — perhaps 25 square miles or 65 square kilometers of coverage, which is similar in range to a cell-phone zone. For line of sight service, WiMAX uses higher frequencies: 66 GHz. Because higher frequencies enable less interference and lot more bandwidth, through the stronger line-of-sight antennas, WiMAX achieves a range of a 30-mile radius (50 kilometers) — perhaps 2,800 square miles or 9,300 square kilometers of coverage.

Following are a few positioning applications that make use of WiMAX technology:

- **Locating stolen laptops:** Typically, WiMAX receivers (such as Bluetooth and/or Wi-Fi) are embedded in laptops, and a stolen laptop can be detected and located.

- **People locating:** Share your location with your friends, family members, doctors, lawyers, colleagues, people in a social network, and so on, and see their location (if they enable sharing). Typically, you run a small application in your WiMAX device.

- **Tracking shipments:** The shipments can be tracked not only at the origin or destination but also along the path.

- **Location-based advertising:** WiMAX provides location-targeted banners and advertisements during Internet access.

How WiMAX works

A WiMAX system consists of two parts (see Figure 10-4):

- **A WiMAX tower:** WiMAX towers are similar in concept to cell-phone towers. A single WiMAX tower can provide coverage to a large area — as big as 3,000 square miles (approximately 8,000 square kilometers).

- **WiMAX receivers (tags):** The receiver and antenna could be contained in a small standalone tag enclosure or extension module for computers and laptops. The WiMAX receiver could be built into your cell phone, the way sometimes Wi-Fi is.

The working principle for locating a tag is very similar to how the location is computed in cellular networks. The computation model is based on one of these methods:

- **Network-based:** One or several WiMAX towers make necessary measurements, such as RSSI, TOA, TDOA, and so on, and forward this data to a central location engine that computes the location

- **WiMAX receivers-based:** The measurements are taken by the receiver and it computes the location

- **WiMAX receivers-assisted:** The receiver collects the measurements and forwards to a central location engine that computes the location.

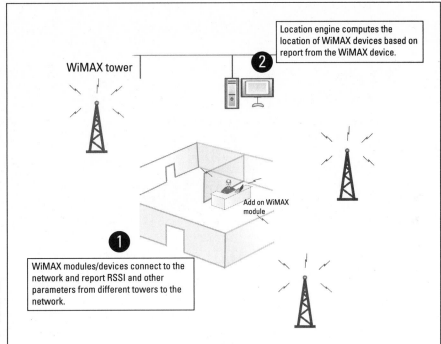

The typical measurements used for locating in WiMAX networks include:

✔ **Received Signal Strength Indicator (RSSI):** Location is computed by trilateration using the RSSI of signals between WiMAX receivers and the WiMAX towers.

✔ **Time difference of arrival (TDOA):** Location is computed by trilateration using time difference of arrival of signal between WiMAX receivers and several WiMAX towers.

✔ **Round Trip Time (RTT):** Trilateration is used to compute location using round trip delay of signal between the WiMAX tower and WiMAX receiver.

Refer to Chapter 2 for more details on how these measurements are used for computing location.

Location accuracy on the order of 50–200 meters can be achieved using WiMAX signals. The accuracy range mostly depends upon the size and density of WiMAX cells.

Pros and cons

Using WiMAX for RTLS offers following key advantages:

✔ **Long range:** The most significant benefit of WiMAX compared with existing wireless technologies is its range. WiMAX has a communication range of up to 30 miles (50 kilometers), or more than 2,800 square miles — enough to blanket a city.

✔ **Urban and indoor locating:** WiMAX can provide location information in urban areas, including indoors.

✔ **Quick initial position fix:** WiMAX gets an initial fix in seconds.

✔ **Standards-based technology:** You don't have to use a particular WiMAX provider to use WiMAX.

✔ **IP-based network:** Because WiMAX is built on IP, it natively runs existing IP-based products, services, and utilities. This system enables much easier and cheaper network monitoring, troubleshooting, and management with standard tools.

✔ **Lower latency:** Because it was designed as a data network from the ground up, WiMAX has a much simpler network topology than cellular networks, which have to add layers and invent new tricks to enable their technology to handle data. This results in lower latency in linking RTLS applications, WiMAX towers, and tags.

WiMAX has the following limitations:

✔ **Not deployed enough:** WiMAX is still in its infancy phase and isn't widely available as of this writing. The clients and tags aren't widely available either.

✔ **High power consumption:** The tags have high battery consumption.

✔ **Low accuracy:** Tracking accuracy is heavily dependent on the size of the WiMAX cell site. Location accuracy is low, on the order of 50–200 meters. Nevertheless, the technology is well suited for situations in which precision better than 100 meters isn't crucial.

✔ **Dependence on carriers.** Some models of locating are dependent upon WiMAX network operators, not all of which may provide this service. It isn't available outside WiMAX networks, which implies you can't locate WiMAX tags in an area where the WiMAX network isn't available.

Understanding TV-signal positioning

Television is one of the most influential forces of our time and although not designed to track location, is a highly suitable option with the potential of widespread deployment for location-based services because of the nature and availability of TV signals — TV signals are almost everywhere (broadcast today across every metropolitan area on Earth), powerful, low and diverse in frequency, and they easily penetrate walls, automobiles, and city buildings.

Following are a few positioning applications that can make use of TV signal–based technology:

✔ **Femtocell synchronization:** Femtocells (discussed in the earlier section, "Improvising a cellular RTLS") are typically deployed in residential or small business environments and require precise location information for their basic operation (for frequency stability and timing synchronization with the cellular network). And by adding a TV signal–based tag to femtocells, the location information of femtocells can be computed automatically. This removes dependence upon the end user to provide the exact coordinates.

✔ **Campus panic alert:** A panic alert enables students or teachers to summon assistance in any part of the school (or university) campus by carrying a TV signal–based call-button tag. There are many areas in a typical campus where it isn't possible to install infrastructure (location sensors) for locating and there isn't GPS coverage either.

✔ **Channel audience analysis:** By integrating the TV-signal tags into televisions, marketers can determine the audience of a specific channel based on the TV's location and use the statistical data for marketing and other purposes.

✔ **Weapon locating:** Attaching TV-signal tags to weapons enables detection and locating of weapons in indoor but dangerous places where it is typically difficult to go.

Television broadcasting frequencies

Broadcast terrestrial (ground-based) television is transmitted on various bands or frequencies. These transmission bands vary by country and are usually set by the nation's spectrum regulation body. For example, in North America, the FCC has allocated each TV channel to a bandwidth of 6 MHz; channels 2–6 use 54–88 MHz), channels 7–13 use 174–216 MHz, and channels 14–83 use 470–890 MHz. Within the 6 MHz space for each channel is a video carrier, a color carrier, and an audio carrier.

Note that the frequencies used for cable TV and satellite television channels are different, and digital television uses the same frequencies as analog. Digital TV, however, requires less bandwidth (more along the lines of 2 MHz).

How television networks can be used for an RTLS

Ground-based television networks consist of a series of TV towers (also known as base stations or broadcast towers). Each tower broadcasts one or more channels. Each channel is transmitted on its own frequency, which can be tuned in and received by your TV set or TV tuner card. Each channel also has timing information. See Figure 10-5 for an example.

The working principle for using this network for an RTLS is as follows:

1. The tag contains a *TV tuner* — a baseband TV measurement module that can extract timing and a back-end network interface (such as Wi-Fi).

2. The tag listens to the channels and forwards the timing information about these channels to the central location engine using the back-end network interface.

3. By using trilateration with knowledge of distance from three or more towers, the location engine is able to compute the position of the tag:

   • Location of towers that broadcast the channel is well-known because each channel uniquely identifies its source tower in an area.

   • The time for signal to travel (from the tower to the tag) can be converted into distance by multiplying it with speed of light.

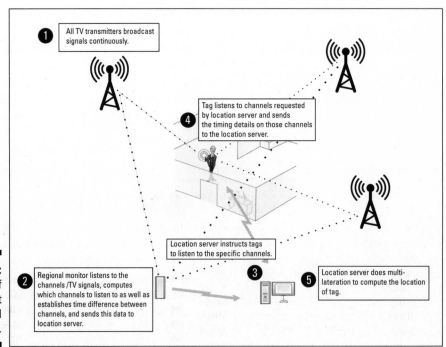

Figure 10-5: Example of deployment of TV signal positioning.

① All TV transmitters broadcast signals continuously.

④ Tag listens to channels requested by location server and sends the timing details on those channels to the location server.

Location server instructs tags to listen to the specific channels.

② Regional monitor listens to the channels /TV signals, computes which channels to listen to as well as establishes time difference between channels, and sends this data to location server.

③

⑤ Location server does multi-lateration to compute the location of tag.

4. Because TV transmitters typically are equipped with clock-reference oscillators that can't match the quality of GPS satellite clocks, the clocks of different TV stations are typically unsynchronized either to each other or a common time reference such as a GPS. For this reason, an external system (called a *signal monitor*) is needed. The signal monitor examines the clock characteristics of different channels to compute correction data. The signal monitor then forwards the correction data to the location engine that can use it for computing location.

To enable a tag to compute its position autonomously (that is, without using an external location engine), the aiding data currently sent from the signal monitor to the location engine could be forwarded to the tag directly or to pseudo-TV transmitters synchronized to GPS time (and the tag uses the signals from these pseudo transmitters).

A typical location accuracy on the order of 5–50 meters can be achieved using TV signals. The accuracy mostly depends upon the density of TV towers.

Pros and cons

TV-signal positioning enjoys fundamental advantages over GPS and A-GPS. These advantages are based on the laws of physics, and their combined effect is compelling:

- **High redundancy:** The broadcast TV infrastructure is distributed, robust, and comprised of transmitters that are highly correlated with population density and broadband penetration. Further, TV transmitters are distributed, supplied with backup power at the studio and hilltop, and are highly resistant to disaster.

- **Better urban locating:** Fading effects, building attenuation caused by building materials, are significantly better for TV than for GPS because

  - TV signals enjoy a substantial power advantage over GPS signals. Not only are terrestrial TV broadcast signals typically 1 megawatt effective radiated power (ERP) compared with 500 watts ERP for GPS satellite signals, but the TV towers are also typically located every 10 miles. The result is that the signal is much stronger when received by the tags in comparison to the signal strength of satellite as received by GPS receivers.

  - Unlike GPS signals (that are transmitted at frequencies of 1.2276 GHz–1.575 GHz), TV signals are broadcast at much lower frequencies (50–800 MHz) and aren't as attenuated by buildings and other man-made structures.

- **Frequency diversity:** Because each TV tower typically broadcasts more than one channel, the receivers have a better chance of acquiring a signal, given that these channels are broadcast at different frequencies.

✔ **Better mitigation of multipath:** Because the bandwidth of broadcast television ranges from 6–8 MHz, in contrast with GPS satellite signal that has a bandwidth of 1 MHz, you can better resolve *multipath* (the reflected signals that characterize urban and indoor environments).

✔ **Resistant to jamming:** A TV signal–based solution is distributed and therefore resistant to jamming.

✔ **Limited atmospheric effects:** Unlike GPS, the TV towers are deployed on ground and so aren't limited by atmospheric effects.

✔ **No ephemeris errors:** Unlike GPS, the TV towers are always at fixed locations and so don't have ephemeris errors. (Ephemeris is discussed in the earlier section, "Using satellite navigation systems.")

Locating using TV signals also has several limitations:

✔ **Need for signal-monitors:** Except for mobile TV networks that are synchronized with GPS, signal monitors are required for aiding data.

✔ **High power consumption:** The TV tuners are power-hungry.

✔ **High cost:** Although the TV tuners aren't as costly for deployment as mobile tags, these tags also need a back-end network radio and corresponding circuitry. The cost is typically lower than other competing wide area network technologies, however.

✔ **No global standardization:** Because different TV standards are used in different countries, the tags need to be different in different countries.

Like TV signals, it is possible to use other signals, such as FM-radio signals, for real-time locating. And like TV signals, positioning synchronization of timing between signals transmitted from different towers is required for FM-radio signals.

Mobile TV signal advantage

Mobile TV is broadcast and delivered in many ways. You can broadcast live TV to mobile devices such as cell phones via satellite, terrestrial towers, or Wi-Fi or WiMAX networks. Many mobile TV signals such as DVB-H (digital video broadcasting for handhelds) or T-DMB (terrestrial digital multimedia broadcasting) are synchronized to GPS, so position can be calculated by the tag without needing any aiding data.

Chapter 11

Precision Locating with Local Area Coverage

....

In This Chapter

▶ Understanding local area coverage

▶ Exploring the underlying technologies

....

*L*ocation is a key organizing principle in the bricks-and-mortar retail world. It's part of the business psyche. Businesses know that it's important to place specific things in specific locations to increase sales. You can see magazines and candies typically placed near checkout stands, for example. And, the commercial success of GPS (Global Positioning System) and the increase in terrorism have spurred everyone's interest — not just the retailers — to automatically locate anything or anybody in real time, not just outdoors but indoors also.

In Chapter 10, I discuss locating technologies, such as GPS, that support *wide area coverage* — locating assets or people in areas spanning large geographic areas, such as cities, countries, and the entire world. As I mention in Chapter 10, although wide area coverage implies locating tags in outdoor, urban, or even indoor environments, typically the technologies that support wide area coverage need some assistance (or require use of precision-locating technologies that support local area coverage) to provide high accuracy in local areas, such as within a room, building, campus, or specific metropolitan area.

In this chapter, I provide an overview of the most commonly used technologies that provide precision locating in local areas, including a description of the strengths as well as the vulnerabilities of each technology.

Understanding Precision

The model of reporting the location for technologies supporting local area coverage is no different from the technologies supporting wide area coverage. Wide area coverage is reported in terms of distance as the *accuracy level* and a percentage as the *precision:*

✔ **Accuracy level:** Accuracy level (or *grain size*) is reported in terms of distance to indicate how far the estimated position could be from the real position.

✔ **Precision:** Precision is reported in terms of a percentage, and it indicates how often you can expect to get that accuracy level.

For example, some Ultra Wideband (UWB)–based RTLS systems (which I discuss later in this chapter) can locate tag positions within 12 inches for approximately 95 percent of measurements. The distance — 12 inches in this example — denotes the accuracy of the position the UWB can provide, and the percentage — 95 percent in this example — denotes precision.

This precision accuracy is often used to determine the locating system's applicability for a particular application. Motion-capture installations may support computer animation features, centimeter-level spatial positioning, and precise temporal resolution, but most applications don't require this level of accuracy. A personal location system for a home or office application might need enough accuracy to answer the query, "Which room was I in around noon?" but not "Where, to the nearest cubic inch, was my left thumb at 12:00:46 p.m.?"

Understanding the Underlying Technologies

Along with the increase in users and in the diversity of users that require precise locating identification in local areas, positioning technologies have received increased attention from various authors, scientists, students, universities, businesses, organizations, and governments. Today, precision locating in local areas can be achieved with many different RTLS technologies that have different physical, operational, performance, and cost characteristics. Precision locating can be achieved with sound, light, camera vision, and radio frequency waves, for example.

Table 11-1 provides a list of the technologies discussed in this chapter, along with the typical level of positioning accuracy achieved with those technologies.

Table 11-1	Typical Accuracy
RTLS Technology	*Typical Accuracy of Location*
Dead reckoning	A few cm
Wi-Fi	1–5 m
Bluetooth	2 m
Passive RFID	1 m
Active RFID	1 m–3 m
Ultra Wideband	30 cm–1 m
ZigBee	1 m
Computer vision	A few cm (depends upon the application)
Acoustic locating	25 m
Building illumination	1–3 m

Technological evolution is a process in which the emergence of the first technology results in a newer species of the same technology or newer technologies. What this means is that your vendor may offer you a slight variation, a combination, or a completely new technology that isn't described in this chapter. Understand how your vendor makes use of the specific technology because that can affect the real-time locating solution's capacity to locate.

The following sections describe the most commonly used technologies that support local area coverage. Note that ultrasound and infrared can be used for precision locating as well. These technologies are discussed in Chapter 9.

Estimating position with dead reckoning

Dead reckoning (DR) is the process of estimating your current position based on a previously determined position, or *fix*. Through this process, you estimate your position based on known speed, elapsed time, and course.

Although traditional methods of DR are no longer considered primary in most applications, DR was used by Christopher Columbus for discovering the new world, and modern inertial navigation systems, which also depend upon DR, are widely used.

In an *inertial navigation system* (also referred to as an *inertial guidance system* or an *inertial reference platform*), tags use sensors to continuously track their position, orientation, and *velocity* (direction and movement speed) and then use this information to compute their precise location as it relates to their start position. Because tags are self-contained and require no location sensors for computing location, DR is an attractive choice for applications when it's difficult to install location sensors. Following are some sample applications that use DR:

- ✔ **Miner safety:** These applications give rescue personnel the most accurate and up-to-date locations of trapped personnel, and also track personnel as they move away from dangerous conditions. Tags are carried by miners and rescue personnel; the tags typically communicate their location to the central location engine/application with a back-end wireless network.

- ✔ **Airplane location:** These applications are used as supplements to Global Positioning System (GPS) locating. These tags are included in the GPS receivers, and whenever GPS loses track of satellites (see Chapter 10 for more on GPS-based locating), the airplane location is updated with the information derived from these tags and the last GPS-reported position.

- ✔ **Firefighter safety:** These applications provide the locations of firefighters when they enter and exit fire areas. The tags are worn by the firefighters; the tags continuously update their location to the central location engine/application with a back-end wireless network.

How dead reckoning works

An inertial navigation system has two main components:

- ✔ **Tags:** Tags have inertial sensing instruments that enable them to detect movement or rotation. Sensing instruments that are typically used in tags include accelerometers, gyros, magnetometers, and barometric altimeters, as shown in Figure 11-1:

  - *Accelerometers:* An accelerometer is a device that measures non-gravitational accelerations. Because mathematically you can use acceleration to deduce the distance change, if you know the direction of travel, you can deduce the tag's current position.

  - *Gyros:* A gyro is a device that measures orientation. By knowing this orientation, you can determine the angle through which the tag has been rotated.

  - *Magnetometers:* A magnetometer measures the strength and/or direction of a magnetic field. You can use this value for calculating the tag estimation with respect to the starting position relative to Earth's magnetic poles. For example, the tag is currently 45 degrees (clockwise) from true north (or magnetic north).

- *Barometric altimeters:* A barometric altimeter (or a *pressure altimeter*) measures the current atmospheric pressure directly as an altitude above mean sea level. You can use the reading to determine a tag's elevation.

Typically, tags have a back-end network interface (such as Wi-Fi) to communicate their locations to the location engine, middleware, and application software.

✔ **Tag start-position initializer:** Because using sensing instruments in a tag can determine only the change in position, the tag's starting position needs to be initialized. For example, when a miner enters the mine, the tag is initialized to indicate the miner's exact coordinates. This way when the miner moves around, the tag always knows the current position of the miner. Typically, this is done by software that informs the tag of the starting position by using the tag's back-end network interface. Another way to establish the starting position is to integrate a locating interface, such as a GPS receiver, into the tag. The GPS can auto-initialize the tag. Before a police officer enters a building, for example, his or her tag can be initialized with the starting position by using GPS, and later positions of the police officer in the building can be estimated based on dead reckoning.

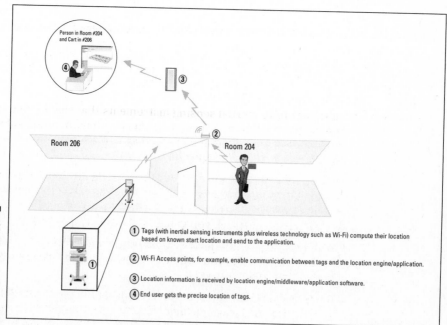

Figure 11-1: Inertial navigation — accelerometers and gyros.

Person in Room #204 and Cart in #206

④

③

②

Room 206

Room 204

① Tags (with inertial sensing instruments plus wireless technology such as Wi-Fi) compute their location based on known start location and send to the application.

② Wi-Fi Access points, for example, enable communication between tags and the location engine/application.

③ Location information is received by location engine/middleware/application software.

④ End user gets the precise location of tags.

Dead reckoning pros and cons

An inertial navigation system enjoys several advantages over other RTLS technologies:

- ✔ **Needs no infrastructure:** The tags require no external input for computing location. This feature is well suited for applications used in spaces that have no existing infrastructure, such as locating firefighters when they go inside a building.

- ✔ **High level of accuracy:** The accuracy of such systems is very high — on the order of centimeters.

- ✔ **Instantaneous output:** A tag always knows its location. No network latency is involved.

- ✔ **Private and secure:** The inertial navigation system was initially developed for the military, so it's accurate, reliable, and not susceptible to signal jamming or erroneous signal transmission. Also, only the tag knows its location, so malicious hackers can't eavesdrop or block the tag from computing its location.

The inertial navigation system also suffers several disadvantages compared with other RTLS technologies:

- ✔ **Drift:** Even the small errors in velocity and direction reported by accelerometers and gyros have the largest impact because each new position is a function of the last position. These errors happen for many reasons, such as vibration, constant change in acceleration or direction of tags, and so on.

- ✔ **High power use:** The sensors consume power to keep the platform aligned with the navigational frame; hence, the life of batteries in tags can be limited. The sensors can be very useful, however, in applications in which tags are put on and removed at entry and exit points, such as personnel tracking in a mine, firefighter tracking, and so on.

- ✔ **Expensive maintenance:** Because the tags are the most sophisticated pieces of hardware involved, all tags require regular maintenance by certified personnel, and repair is expensive as well.

- ✔ **Need for precise start-up position:** Because the tags can only determine the change in position, the tag's starting position needs to be initialized precisely. Any initial entry errors have a direct impact on the current position determined by tags.

Using Wi-Fi for an RTLS

Wi-Fi (short for *wireless fidelity*) is a technology used around the world to connect to the Internet or any network without wires or cables. The technology is nearly ubiquitous: Places from entire cities to coffee shops and fast-food restaurants to offices and homes offer Wi-Fi; all sorts of devices from computers to cell phones to TVs and console game systems use Wi-Fi. When you use Wi-Fi technology, your device connects wirelessly to a Wi-Fi *access point* (also called a *wireless access point* or a *wireless router*) that provides the link to the Internet or other type of network.

Wi-Fi is also referred to as 802.11 networking. The 802.11 designation comes from the Institute of Electrical and Electronics Engineers (IEEE).

Although, the traditional role of Wi-Fi is to provide convenient and widely available wireless network access, the maturing and widespread acceptance of Wi-Fi has begun a flurry of activity regarding an RTLS using Wi-Fi. Businesses are already deploying Wi-Fi for internal use or for guest access to the Internet and by using it for the RTLS, they can extend the benefits of the deployment to areas such as audit, compliance, security, and asset management, and in general can improve efficiency throughout the enterprise.

How a Wi-Fi RTLS works

Many Wi-Fi–based location systems are commercially available or in the research phase. The basic principles used by most Wi-Fi RTLS are Received Signal Strength Indicator (RSSI) and Time Difference of Arrival (TDOA), both of which are covered in the following sections.

A Wi-Fi RTLS can locate laptops, personal digital assistants (PDAs), and other devices that have embedded Wi-Fi radios or tags. I also use the term *Wi-Fi tag* to refer to devices with built-in Wi-Fi radios.

RSSI-based Wi-Fi RTLS

Received Signal Strength Indication (RSSI) is a measurement of the power present in the signal received by Wi-Fi tags or Wi-Fi access points from one another. Because the power levels at the start of the signal transmissions are well known and the power drop in signals in open space as well as through different media is well defined, RSSI can be used to estimate the distance between Wi-Fi tags and Wi-Fi access points.

Wi-Fi includes IEEE 802.11 standards, including 802.11a, 802.11b, 802.11g, and 802.11n.

✔ **802.11a** operates at 5.8 GHz spectrum with the Orthogonal Frequency Division Multiplexing (OFDM) modulation technique. The 5.8 GHz spectrum is further divided into many channels (the number of channels depends upon the country regulations).

✔ **802.11b** operates at 2.4 GHz spectrum with the Direct Sequence Spread Spectrum (DSSS) modulation technique. The 2.4 GHz spectrum is further divided into 14 overlapping (3 non-overlapping) staggered 20 MHz wireless carrier channels (channels 1–14) whose center frequencies are 5 MHz apart. In simple terms, DSSS is a spread spectrum technology (where the carrier signals occur over the full bandwidth — the spectrum of a device's transmitting frequency) in which the signal is encoded using a pseudo-noise code sequence.

✔ **802.11g** operates at 2.4 GHz spectrum like 802.11b, but uses the same OFDM-based transmission scheme as 802.11a. 802.11g is interoperable with 802.11b.

✔ **802.11n** operates at both the 2.4 GHz and 5 GHz spectrums and builds upon earlier 802.11 standards by adding Multiple-Input Multiple-Output (MIMO). 802.11n provides ways of ensuring coexistence among the legacies 802.11a, b, and g and the new devices. 802.11n is expected to be finalized in November 2009.

The channels that are available for use in a particular country differ according to the regulations of that country. In the United States, for example, FCC (Federal Communications Commission) regulations allow only channels 1–11 to be used in 2.4 GHz spectrum; in Europe, channels 1–13 are licensed for 802.11b operation (with 1, 5, 9, and 13 usually deployed); and in Japan, all 14 channels are licensed for 802.11b operation.

Typical Wi-Fi access point transmission power is 100 milliwatts.

Locating Wi-Fi tags with RSSI is implemented in one of the following ways:

✔ **Network-based:** As shown in Figure 11-2, in network-based implementation, the working principle is

1. A Wi-Fi tag does the active signal transmission. A Wi-Fi tag sends Wi-Fi *probe requests* (typically, Wi-Fi clients send probe requests whenever they need to determine which access points are within range), proprietary Wi-Fi messages, or some other Wi-Fi messages.

2. One or several access points that receive this signal make the measurements of RSSI and forward this information to a location engine. The access points are connected to the location engine by a wired or wireless network.

3. Based upon these RSSI values and access point locations, the location engine computes the tag's location with one of the techniques I describe later in this section.

✔ **Client-based:** As shown in Figure 11-3, in client-based implementation, the working principle is

1. A Wi-Fi tag listens for *beacons* (Wi-Fi access points send beacons periodically to announce their presence and provide parameters for configuration for Wi-Fi clients in range) in the air and records the RSSI values from various access points.

2. The Wi-Fi tag forwards these RSSI value/access point pairs to the location engine, typically with the Wi-Fi back-end network.

3. Based upon these RSSI values and access point locations, the location engine computes the location with one of the techniques I describe later in this section.

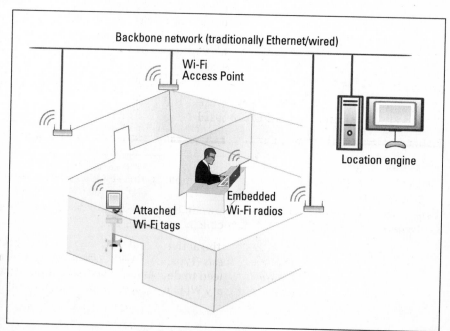

Backbone network (traditionally Ethernet/wired)

Wi-Fi
Access Point

Location engine

Embedded
Wi-Fi radios

Attached
Wi-Fi tags

Figure 11-2:
A Wi-Fi
RTLS.

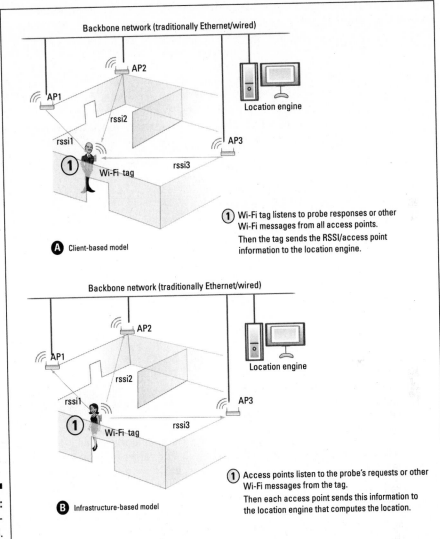

Figure 11-3:
An RSSI-based RTLS.

The location engine uses the RSSI values and one of the following methods to determine the location of a Wi-Fi tag:

✔ **Nearest access point:** In this method, the location engine returns the location of the nearest Wi-Fi access point (highest RSSI) as the location of the Wi-Fi tag. This method is the simplest but least accurate way to locate a Wi-Fi tag. If you have an access point deployed every 100 feet, for example, this method narrows the location of a Wi-Fi tag to an area of only 10,000 square feet. For this reason, this method is sometimes called *presence detection* or *presence-based locating*. Depending on your application's location-accuracy needs, this method may be adequate, but in most cases isn't sufficient.

✔ **Trilateration:** In this method, the location engine uses trilateration (I discuss this in Chapter 2) to determine the most likely location of a Wi-Fi tag. The basic methodology is that if the distance *d* from various access points to the Wi-Fi tag can be measured, a circle with radius *d* can be drawn with the center at each access point. The most likely location of the device is where all these circles intersect (see Figure 11-4).

RF signal attenuation

As the signal leaves its source, it *attenuates,* or the power of the signal drops. The signal attenuates in free space, depending on the absorption characteristics of the objects in its transmission path (the objects in the signal's path are also dubbed *barriers*).

The free-space propagation loss in dB is given by the formula:

Path loss (approx) = $-38 + 20 \times \log_{10}(f) + 10 \times n \times \log_{10}(d)$

In this formula, *f* is the transmission frequency in MHz; *d* is the distance specified in feet; and *n* is the path-loss constant depending upon the environment. The value of *n* is 2 for free space, 2.5–4 in most indoor environments, and so on. For example, the free-space loss for 2.4 GHz at 100 feet from the transmitter is about 70 dB.

Based on the transmission frequency, the thickness, and the specific type of material used, different materials absorb radio frequency (RF) energy and cause RF attenuation. For example, a steel fire-exit door causes an attenuation of 13–19 dB for a 2.4 GHz signal and 25–32 dB for a 5 GHz signal; interior drywall causes an attenuation of 3–4 dB for a 2.4 GHz signal and 3–5 dB for a 5 GHz signal.

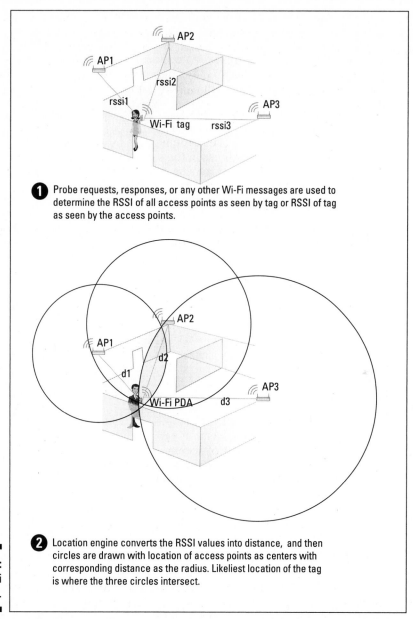

1 Probe requests, responses, or any other Wi-Fi messages are used to determine the RSSI of all access points as seen by tag or RSSI of tag as seen by the access points.

2 Location engine converts the RSSI values into distance, and then circles are drawn with location of access points as centers with corresponding distance as the radius. Likeliest location of the tag is where the three circles intersect.

Figure 11-4:
Wi-Fi
trilateration.

To find the distance from each access point, the RSSI values of messages between access points and the Wi-Fi tag are converted to distance via a signal propagation model.

Because RSSI attenuates exponentially with distance, if the tags are closer to access points, RSSI can be almost linearly correlated with distance. This means that location accuracy is directly proportional to the density of access points. The more access points you deploy, the greater the location accuracy. Conversely, the fewer access points you deploy, the less accurate the measurements are.

✔ **Fingerprint match (scene analysis):** In this method, also known as *calibration* or *training,* actual measurements of RSSI values (as seen by real Wi-Fi tags in the real environment) from various access points are recorded at different places in the facility and stored in a database. This fingerprint database may be comprehensive, such as collected at a very large number of points in the facility, or it may just include measurements from known pathways, which is more commonly used commercially. The location engine estimates the location of a Wi-Fi tag by comparing RSSI observations at the current location with observations in a database. The outcome of the search and matching algorithms is the likeliest location of the tag. ***Note:*** These databases also need to be rebuilt often due to environmental changes.

✔ **Calibration (sample) tags:** In this method, various Wi-Fi tags, called *calibration tags,* are placed at well-known fixed locations throughout the facility. The location engine estimates the location of a Wi-Fi tag by comparing the RSSI observations at its current location with observations from various calibration tags. The calibration tags are typically used in conjunction with the fingerprint-match models or trilateration-based models to enhance location accuracy.

✔ **Landmark tags:** In this method, various tags, called *landmark tags,* are placed at fixed locations throughout the facility. The location engine estimates the Wi-Fi tag location based on RSSI values from the landmark tags and the RSSI values of access points. Because the relative distances between Wi-Fi tags and landmark tags are shorter than the conventional distances between Wi-Fi tags and Wi-Fi access points, obstacles are less of a problem during positioning. Using landmark tags enables the location engine to estimate location with high accuracy, even in environments that don't have enough access points deployed.

TDOA-based Wi-Fi RTLS

In a TDOA-based Wi-Fi RTLS(I discuss TDOA in Chapter 2), location is computed by trilateration using access point locations and the time difference between the signal received by multiple access points and the Wi-Fi tag. The working principle of a TDOA-based Wi-Fi RTLS can be described as

1. A Wi-Fi tag does active signal transmissions. A Wi-Fi tag sends Wi-Fi *probe requests* (typically Wi-Fi clients send probe requests whenever they need to determine which access points are within range), proprietary Wi-Fi messages, or some other Wi-Fi messages.

2. Each access point records the time when it receives the signal and then forwards that information in the form of a timestamp to the location engine.

3. The location engine uses the position of access points along with the received signal's time difference between each of the access points. Mathematically, for trilateration purposes, the tag position computed by TDOA is the intersection of hyperbolas (hyperboloids in 3D).

To measure time differences in the signals received by multiple access points from a Wi-Fi tag, you may need specialized access points or hardware. This makes the Wi-Fi network proprietary from this perspective. You may not be able to mix and match Wi-Fi access points from various vendors. You'll also need to have the clocks of the access points synchronized precisely for TDOA to work. Many times wires or cables are used to achieve time synchronization between access points.

Because TDOA estimates time based on the shortest received path (usually, a direct path), this method is most effective in a multipath environment. Accuracy is affected by latency in receiver response (which may be due to a processing queue at the receiver) but isn't affected by the distance between tags and access points.

Wi-Fi RTLS pros and cons

Wi-Fi RTLS products enjoy several advantages over other RTLS technologies:

✔ **Standards-based:** The Wi-Fi RTLS is a standards-based approach that translates to more choices and interoperability features in a multiple-vendor solution, which usually results in lower prices.

Although Wi-Fi is a standard, the implementation of a Wi-Fi RTLS could be done in proprietary ways, and RTLS implementation may not work with W-Fi access points from all vendors. You must verify this with your Wi-Fi RTLS vendor.

✔ **Lower cost:** A Wi-Fi RSSI-based RTLS tends to be much less expensive than other technologies because it makes use of Wi-Fi access points installed for data network/Internet access.

✔ **Less training:** Because many Wi-Fi networks are already installed, you don't have to teach personnel how to manage such networks.

✔ **Multipurpose network:** Wi-Fi is a multipurpose network — you can use it for data, video, voice, and so on.

✔ **Simplified troubleshooting:** Managing multiple, incompatible proprietary systems usually is difficult because the systems require separate consoles and additional training. If the system breaks down, vendors may blame other vendors' equipment for the problems. Using a single technology in the network simplifies troubleshooting.

✔ **Tracking without tags:** With growing categories of devices using built-in Wi-Fi, organizations can track other assets (such as Wi-Fi–enabled PDAs, mobile phones, and laptops) without attaching tags to them.

Wi-Fi RTLS products also suffer several disadvantages compared with other RTLS technologies:

✔ **Burden on Wi-Fi:** Although the burden is minimal, a Wi-Fi RTLS does add some burden on the Wi-Fi network via data traffic from an RTLS. The burden depends on the number of tags involved and how often locations are updated.

✔ **Limited accuracy of existing infrastructure:** Existing Wi-Fi infrastructure typically isn't installed for high accurate location information. Most companies have to reposition their access points or add a significant number of new access points to ensure dense-enough coverage to locate equipment and people.

✔ **Scalability limits:** The scalability of a Wi-Fi locating system depends on the performance of the server (location engine) that calculates the position.

✔ **Limits of RSSI:** RSSI is affected by various environmental features, such as obstacles, multipath fading, temperature and humidity variations, opening and closing of doors, furniture relocation, and the presence and mobility of human beings. Consequently, the value of RSSI recorded by the receiver could be much higher or lower than the anticipated value. Also, RSSI not only differs from the theoretical value but also fluctuates at the same position. Some factors affecting RSSI are described below:

  • *Multipath fading:* Multipath fading occurs when the receiver sees the overlapping of multiple copies of the transmitted signal, each traversing a different path. This effect is caused when items such as people, chairs, desks, and other objects get in the way and cause the signal to bounce in different directions. A portion of the signal may go directly to the destination; another part may bounce from a chair to the ceiling and then to the destination. As a result, some of the signal encounters delay and travels a longer path to the receiver (see Figure 11-5). As a result, the RSSI value received by the receiver isn't necessarily the value that can be used for locating.

  • *Time delay:* An anomaly that goes hand in hand with multipath fading, *time delay* is the amount of timing variation among different radio frequency (RF) signals. Time delay can cause phase and polarization changes as well as multipath fading. Unlike multipath fading, which affects signal amplitude, time delay adversely affects the receiver's ability to decode signals due to distortion.

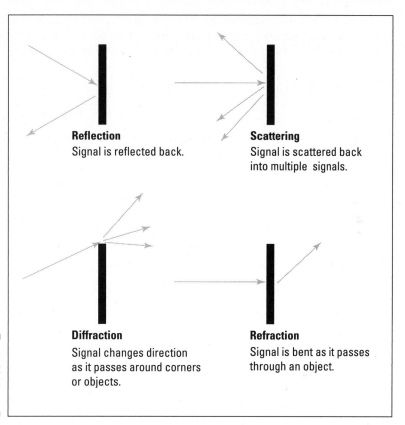

Reflection
Signal is reflected back.

Scattering
Signal is scattered back into multiple signals.

Diffraction
Signal changes direction as it passes around corners or objects.

Refraction
Signal is bent as it passes through an object.

Figure 11-5:
What happens to signals.

- *Aliasing:* Another problem in using RSSI is *aliasing,* which occurs when several locations receive the same signal strength of an access point. Even worse, due to variations in the signal strength caused by obstacles, the two locations need not be the same distance from the access point. This partly explains why a simple trilateration of the position via the signal strength doesn't lead to an accurate estimate because there are multiple positions with the same RSSI characteristics.

- *Doppler effect:* RF Doppler effects are created in two ways: relative motion between the transmitter and receiver, and RF non–line-of-sight conditions that alter the relative motion of the RF signals themselves and can prevent the receiver from decoding signals from a Wi-Fi tag. In the latter case, locating is done without data from one access point, and the results can get distorted.

✔ **Difficult to fingerprint:** For a fingerprint-match Wi-Fi RTLS (see the section, "RSSI-based Wi-Fi RTLS," earlier in this chapter), someone has to collect data to see how RSSI looks like in relation to real locations. This method has the following limitations:

 • *Effort in calibration:* Calibration isn't a big deal if you're calibrating for a small office space, but it becomes increasingly complex to calibrate for a large space, such as a 50-story building.

 • *Dynamic environment:* Several environment issues — such as failing access points, structures that block signals, new partitions, and more people — may affect the quality of a fingerprint-match database, so the database must be rebuilt more often, which takes additional time and cost.

✔ **Interference:** Many other types of devices emit in the 2.4 GHz (the ISM [industrial, scientific, medical] unlicensed) band. These devices include microwave ovens, cordless phones, Bluetooth devices, wireless video cameras, outdoor microwave links, wireless game controllers, ZigBee devices, fluorescent lights, WiMAX, and so on.

✔ **Security:** Many of the network attacks discussed in Chapter 15 could be executed against a Wi-Fi RTLS, including jamming, blocking, man-in-the-middle, and denial-of-service attacks. In addition, you need the ability to locate *rogue access points* or *rogue Wi-Fi clients.* Rogue Wi-Fi clients or access points are the ones that aren't authorized and are trying to intercept traffic or steal passwords, steal security parameters, and launch attacks.

Locating with Bluetooth

Bluetooth is a wireless networking standard designed for low power consumption and communication in a personal area networking (PAN) environment. Many cell phones are Bluetooth-enabled, for example. The technology wasn't designed to perform locating, but Bluetooth devices are ideal for locating because they contain a mechanism to identify their neighbors and communicate with other devices in the area.

How Bluetooth is used for precision locating

The working principle of a Bluetooth RTLS can be described as

1. Bluetooth access points are similar to Wi-Fi access points but installed more sparsely (typically, 10–15 meters apart).

2. Bluetooth tags could be standalone tags; built-in features of computers, PDAs, cell phones, and other devices; or expansion cards that can be added to other devices.

3. To compute location, the location engine instructs all the Bluetooth access points to either find all nearby tags or find a specific tag:

  • *Find all nearby tags:* To find all nearby tags, use the Bluetooth *inquiry procedure.* The inquiry procedure enables a Bluetooth device to discover which devices are in range and then determine the addresses and clocks for the devices. An inquiry process typically takes 5–10 seconds and gives the inquirer (the access point) the IDs of all the Bluetooth devices (tags) within its RF range.

  • *Find a specific tag:* To find a specific tag, use the Bluetooth *paging procedure.* By making use of the paging procedure, the inquirer (access point) can *page* (set up a connection with) one or more of its discovered neighbors (tags). A paging process typically takes 1–2 seconds. This mechanism is faster than inquiry procedure but requires a previous knowledge of the tag's ID and clock information (as obtained by the inquiry procedure).

4. The Bluetooth access points report the tag's RSSI to the location engine.

5. The location engine makes use of proximity, trilateration, or fingerprinting to determine the location. Because Bluetooth is RF-based and operates in the same frequencies as Wi-Fi, for detailed descriptions of these methods, see the earlier section, "RSSI-based Wi-Fi RTLS."

Bluetooth frequencies

Bluetooth operates on the 2.4 GHz band, which is the same as Wi-Fi, cordless phones, and various other wireless devices.

Bluetooth makes use of *FHSS, Frequency Hopping Spread Spectrum.* In FHSS, a device will use one of 79 different, randomly chosen frequencies within an assigned range and will frequently change frequencies from one to another. Bluetooth-enabled devices hop frequencies 1,600 times per second. As a result,

more devices can use a portion of the radio spectrum. Bluetooth devices also make use of Adaptive Frequency Hopping (AFH) to avoid crowded frequencies or to avoid channels that have bad quality of wireless signal.

Typical Bluetooth device transmission power is 1 milliwatt for devices with a 1-meter range, 2.5 milliwatts for devices with a 10-meter range, and 100 milliwatts for devices with a 100-meter range.

Bluetooth RTLS pros and cons

Using Bluetooth for an RTLS offers the following advantages:

- ✔ **Standards-based:** A Bluetooth RTLS is a standards-based approach that translates to more choices and interoperability features in a multiple-vendor solution, which usually results in lower prices.

- ✔ **Low power consumption:** Bluetooth is a low-power technology.

- ✔ **Tracking without tags:** With growing categories of devices using built-in Bluetooth, organizations can track other assets (such as Bluetooth-enabled PDAs, mobile phones, and laptops) without attaching tags to them.

- ✔ **High accuracy:** Because Bluetooth access points are placed relatively close to each other, location accuracy is about 2 meters.

- ✔ **Multipurpose network:** Bluetooth infrastructure is a multipurpose network and provides additional services, such as remote monitoring and control, voice, and Internet Protocol (IP) services (although Wi-Fi is more suitable for IP services).

Bluetooth for an RTLS also has some disadvantages:

- ✔ **Short range:** Because Bluetooth range is short, access points need to be installed every 15–20 meters.

- ✔ **Security:** Many network attacks, which I discuss in Chapter 15, could be executed against a Bluetooth RTLS, including jamming, blocking, and denial-of-service attacks.

- ✔ **Slow response:** Due to the Bluetooth inquiry process, the positioning delay is about 15–30 seconds.

Locating using passive RFID

Because passive RFID tags have no batteries and are typically very small in size, they're an attractive choice for precision locating. Although there are many passive RFID (radio frequency identification) technologies (which I discuss in Chapter 8), the focus of precision locating is on EPC Gen 2 tags (*EPCglobal UHF Class 1 Generation 2,* a standard approved for passive RFID in 2004).

Many real-time location systems that use EPC Gen 2 tags are commercially available or in the research phase. EPC Gen 2 tag locating is implemented with the *tightly coupled model* or *distributed model,* both of which are covered in the following sections.

Tightly coupled model

This is the conventional model in which EPC Gen 2 tags are read by RFID interrogators (or *readers*). The working principle of this model can be described as

1. One or more RFID interrogators are installed in the area where precision locating is desired. Each interrogator has one or more antenna connected to it. All antennas are placed such that at the same location, a tag can be read by multiple antennas. ***Note:*** The passive RFID reader's range is directly proportional to the size of the reader antenna or antenna on the tag; the range typically is 20 feet, but some large passive RFID tags can be read up to 50 feet.

2. The location engine instructs the readers to read specific tags or all tags in the area.

3. Readers read tags and record the RSSI of tags read. Readers then forward the tag data and RSSI data to the location engine.

   Not all passive RFID readers from various vendors report RSSI. Verify this feature with your vendor.

4. The location engine computes the tag locations by trilateration with the antenna location and RSSI values.

Gen 2 protocol anticipates situations where there are several readers simultaneously reading the same tag. Up to four separate readings of the same tag can be undertaken simultaneously without interference — and without having to wait for any one reader to complete its read.

Tightly coupled model pros and cons

The tightly coupled model for locating using EPC Gen 2 tags makes an excellent solution for precision locating of smaller things (when it's extremely important to have a really small tag size) in smaller zones (for example, locating surgical equipment in an operating room).

Here are a few advantages of this model:

- ✔ **Low tag cost:** EPC Gen 2 tags are at very low order costs, a few cents.
- ✔ **Embedding:** Tags can be embedded easily in solid nonmetallic items (labels, pallets, cards, and so on).

✔ **High throughput:** High data throughput and a faster anticollision scheme facilitate higher read rates (typically 1,500 tags per second).

✔ **Good accuracy:** The tags can be located precisely with an accuracy of 1 meter using three or more antennas.

The model has a few disadvantages also:

✔ **Missing global standardization:** No global standards and regulations exist for frequency.

✔ **Environmental tolerance:** UHF (ultra high frequency) RFID perform poorly around liquids and metals.

✔ **Crowded frequency band:** The 860–960 MHz range falls within the ISM (industrial, scientific, medical) band, making it one of the most crowded regions of the spectrum.

✔ **Large number of required readers:** If you want to locate in large areas, install a large number of readers throughout the facility.

✔ **High network traffic:** The system generates significant network traffic because the location engine needs to communicate with all readers in the facility at the same time.

✔ **Extended location-engine capabilities:** The location engine needs to have business logic to produce results in the presence of cross-reads of RFID tags by multiple readers.

Distributed model

In the distributed model, the two activities of RFID readers — providing energy to tags to power-up and reading the tags — are separated. The working principle of this model can be described as

1. Instead of deploying RFID interrogators, multiple transmit devices and one receiver device are installed in the area where precision locating is desired. The purpose of transmitters is to energize (power-up) the tags, and the purpose of receivers is to read the tag signals. The range of transmitters is typically 20–30 feet, and the receiver's range is much larger (200,000 square feet, for example). Typically, the receiver makes use of smart antennas, space-time array techniques, and so on to detect very faint signals across vast distances in an extremely noisy RF environment.

2. The location engine instructs the transmit devices to excite the tags in its range (such as providing the tags energy so that they can respond).

3. Tags transmit their information by backscatter.

4. The RFID receiver reads the resulting tag signals from a distance and forwards this information to the location engine that computes location.

Distributed model pros and cons

The distributed model for locating with EPC Gen 2 tags makes an excellent solution for precision locating in large areas. All the advantages that are listed in the preceding section, "Tightly coupled model pros and cons," apply, and additional advantages include

- **Large areas:** EPC Gen 2 tags can be precisely located in very large areas.
- **Eliminates cross-reads:** Because reads are done at a single point (the receiver) and transmit devices are there only to excite tags to transmit signals, it eliminates cross-reads (which is very common in tightly coupled model-based installations).
- **Good accuracy:** The tags can be precisely located with accuracy of 1 meter with one receiver.

The disadvantages listed in the preceding section, "Tightly coupled model pros and cons," apply for distributed model, in addition to the following:

- **Missing global standardization:** No global standards and regulations exist for frequency.
- **Environmental tolerance:** UHF RFID performs poorly around liquids and metals.
- **Crowded frequency band:** The 860–960 MHz range falls within the ISM band, making it one of the most crowded regions of the spectrum.

Using active RFID

Although the term *active RFID* can be used to describe almost any active tag that uses RF-based wireless technology — UWB, ZigBee, Bluetooth, 900 MHz, Wi-Fi, and even cellular — the term has been traditionally used for closed loop active RFID systems.

The most commonly used frequencies by active RFID-based locating systems are listed in Table 11-2.

Table 11-2	Typical Active RFID	
Class	*Frequency*	*Some Characteristics*
UHF 300 MHz – 1 GHz	303 MHz (302–305 MHz) 315 MHz (314.7–315 MHz) 418 MHz (418.95–418.975 MHz) 433 MHz (433.050–434.780 MHz) 868 MHz (868–868.6 MHz) 915 MHz (902–928 MHz)	Exhibit good penetration through non-conductive materials and non-conductive liquids. Lesser material absorption in comparison to microwave.
Microwave	2.45 GHz (2.4–2.4835 GHz) 5.8 GHz (5.725–5.8775 GHz)	Exhibit good penetration through non-conductive materials but are absorbed by water and water-based solutions. Reflected by metals and other conductive surfaces.

Many applications with active RFID for RTLS have been around for decades. Following are a few RTLS applications that use active RFID-based systems:

✔ **Road toll:** Nonstop road tolling for vehicles with readers deployed at tollbooths and tags in the vehicles.

✔ **Area monitoring:** This includes many applications, such as collecting real-time inventory information within a warehouse, monitoring the security of ocean containers or trailers stored in a yard, monitoring the location of air cargo containers across an air terminal or tarmac, and so on.

✔ **Logistics:** This includes many applications in the field of logistics, such as tracking of luggage, transporting of frozen food and pharmaceutical products, and so on.

✔ **Security:** Access controls for goods and persons.

No single frequency is ideal for all applications, even within a single industry. And, implementations for RTLS products based on the same frequency vary, also.

Because frequencies affect the cost, size, and performance of your RTLS, to select the optimal radio frequency for your specific application need, consider the following guidelines:

- **Lower frequencies *propagate* (or travel) farther than signals at higher frequencies (assuming similar transmitter power levels).** This is in general true for frequencies greater than 100 MHz. Systems less than 100 MHz depend upon inductive coupling (which I describe in Chapter 8), and the coupling range drops sharply with distance. Also, lower frequencies propagate better through crowded environments because that signal can diffract around an object when the wavelength approaches the size of the object (lower frequency means higher wavelength). For example, at 433 MHz the wavelength is 69.28 cm, enabling signals to diffract around obstructions, such as vehicles. At 2.4 GHz, the wavelength is approximately 12.5 cm, and diffraction is very limited with these obstructions, creating areas of limited coverage. However, it's important to note that range is also a factor of power used for transmission as well as shape, size, and directionality of the antenna used.

- **Higher frequencies have higher data transfer rates.** This means that the tags can be read at higher rates and large quantities of data can be sent between tags and receivers in a very short time. In an industrial application, it may be required to process the location of thousands of employees (and visitors, if any) with an accuracy between 1 and 3 meters, every second.

How active RFID is used for an RTLS

Because active RFID-based solutions are mostly proprietary, implementations from different vendors vary. Typical models of implementation with active RFID are either with *actively transmitting tags* or *actively listening tags*.

In cases of actively transmitting tags:

1. Active RFID tags continuously chirp, or send signals at a preset frequency, say every 10 seconds.

2. These signals are received by active RFID readers installed in ceilings or walls.

3. The active RFID readers take measurements, such as Time of Flight (ToF), Received Signal Strength Indication (RSSI), Time of Arrival (TOA), and Angle of Arrival (AOA), of these signals and then forward these measurements to the central location engine.

4. The central location engine computes the location with an applicable technique, such as trilateration, fingerprinting, and so on.

And, in cases of actively listening tags:

1. Active RFID tags periodically wake up (say, every 2 seconds) to listen for any messages from the active RFID readers.

2. Upon receiving signals, these tags do one of the following:

    • Take measurements, such as TOF or RSSI, of these signals and then forward these measurements to the location engine by using a back-end network, or the tags just signal.

    • Send signals that are received by RFID readers that take measurements, TOF, RSSI, TOA, and AOA, of these signals and then forward these measurements to the central location engine.

3. The central location engine computes the location with an applicable technique, such as trilateration, fingerprinting, and so on.

Active RFID RTLS pros and cons

The key advantage of using active RFID for an RTLS is that it offers a business focus. The whole argument for closed loop applications is solving a business problem — quickly and effectively. The other advantages include

✔ **Overlay network:** Because the active RFID RTLS uses a separate infrastructure (location sensors), the existing infrastructure (the network deployed for data, video, voice, and so on) isn't impacted by the RTLS application.

✔ **Innovative solutions:** Because vendors aren't forced to conform to an industry-wide standard, they typically come up with innovative solutions.

The model has also have several disadvantages:

✔ **Missing global standardization:** No global standards and regulations exist for an active RFID-based RTLS.

✔ **Maintenance challenge:** Because the active RFID RTLS uses a separate infrastructure (location sensors), this creates additional maintenance and support overhead for your IT.

✔ **Multiple-vendor interoperability:** These solutions typically don't provide multiple-vendor interoperability. You may have a challenge when moving from one generation of products to the next.

✔ **Supplier dependence:** You're at higher risk with sole-sourced solutions. Standards-based solutions enable you to buy interoperable systems from multiple vendors.

The security aspect of closed loop applications is often debated. Each side makes arguments and then refutes the arguments of the other side. In truth, neither is correct (or both are). See Chapter 15 on RTLS vulnerabilities to understand the security aspect of your RTLS application.

Air interface standards

Although an active RFID-based RTLS is typically a proprietary implementation, there are standards for using an air interface at different frequencies, and there's an ISO (International Organization for Standardization) standard for the RTLS using 2.4GHz:

- ✓ **ISO 18000-1:** Generic Parameters for the Air Interface for Globally Accepted Frequencies

- ✓ **ISO 18000-4:** For 2.45 GHz

- ✓ **ISO 18000-6:** For 860–960 MHz

- ✓ **ISO 18000-7:** For 433 MHz

- ✓ **ISO/IEC 24730:** Defines two air interface protocols and a single application program interface (API) for the RTLS for use in asset management and is intended to allow for compatibility and to encourage interoperability of products for the growing RTLS market. The ISO 24730 system uses tags that use 2.4 GHz, DSSS (Direct Sequence Spread Spectrum). These tags may also include low-frequency (125 kHz) tags to support an optional exciter mode (see Chapter 8) that allows tag location updates based on their proximity to an exciter. The method of location using a 2.4 GHz beacon is generally through trilateration with methodologies, such as TOF, RSSI, Time Difference of Arrival (TDOA) and AOA; it also describes a third interface — On/Off Key (OOK) — that allows a tag to communicate with a handheld reader.

Using ultra-wideband (UWB)

Ultra Wideband (*UWB* or *ultraband*) is any radio technology that has bandwidth exceeding 500 MHz or 20 percent of the arithmetic center frequency, whichever is lower. UWB is a carrierless communication scheme. The early applications of UWB technology were primarily related to radar. Table 11-3 lists the categories of applications approved by the Federal Communications Commission (FCC) for UWB.

Table 11-3	UWB Applications
Application/Class	*Frequency*
Communications and measurement systems	3.1–10.6 GHz
Imaging: ground-penetrating radar, wall, medical imaging	Less than 960 MHz or 3.1–10.6 GHz
Imaging: through walls	Less than 960 MHz or 3.1–10.6 GHz
Imaging: surveillance	1.99–10.6 GHz
Vehicular	24–29 GHz

How UWB works

A UWB-based locating system is very much like any other RTLS except that it uses UWB signals. Here are the parts of a UWB-based locating system and descriptions of how they work:

✔ **Tags:** The tags send UWB pulses, which usually are short and have low repetition rates, typically 1–100 megapulses per second.

✔ **UWB receivers (location sensors):** UWB receivers extract timing (and sometimes angle) information from the UWB signals emitted from tags and send the information to the centralized location engine that computes the location.

Two types of receivers are used, the first relatively complex and the second relatively simple:

- *Coherent receiver:* A coherent receiver counters the effects of multipath fading by using several subreceivers. Coherent receivers offer longer range, more accuracy, and better performance in noisy channel conditions.

To collect energy in an intensely multipath environment, you must use a large number of subreceivers. In addition, you may require a high-speed precision clock. The solution is an expensive one.

- *Non-coherent receiver:* A non-coherent receiver is easy to use, but you experience reduced channel capacity. Depending on the application's sensitivity to accuracy, non-coherent receivers may be acceptable, even though they may have shorter communication range, less accuracy, and inferior performance under certain channel conditions.

✔ **Location engine:** The location engine computes the location based on the tag-specific data received from different UWB receivers. As a UWB RTLS uses short pulses with low repetition rates, the location engine can compute location in a few ways:

- *Angle of Arrival (AOA):* The AOA technique (described in Chapter 2) measures the angle of a signal between a given tag and different UWB receivers to estimate the location.

- *Time Difference of Arrival (TDOA)* or *Time of Arrival (TOA):* Unlike conventional RF, UWB signals have very large bandwidths, which allows for extremely accurate location estimation using the TDOA or TOA approach (both discussed in Chapter 2).

UWB pros and cons

UWB-based positioning enjoys several advantages over other RTLS technologies:

✔ **High accuracy and penetration:** UWB provides very accurate locating. Accuracy can be up to a few centimeters, and results can be produced in three dimensions. Location-aware applications can tell exactly which room a tracked person is in, which chair she's seated in, and which mobile devices are closest to him or her so that the right responses can be made.

✔ **Excellent performance in challenging RF environments:** In highly reflective environments that have high metal content, such as manufacturing plants, UWB maintains excellent performance because of its short low-duty-cycle pulses.

✔ **No interference with other RF systems:** UWB tags can be used in close proximity to other RF signals because they don't cause or suffer from interferences due to the differences in signal types and radio spectrum used.

✔ **Relative immunity to multipath fading:** Because UWB pulses are narrow and occupy the entire UWB bandwidth, UWB isn't susceptible to multipath fading.

But the high accuracy advantages come at a cost:

✔ **Need for line of sight or cables:** Line of sight or timing cables between various UWB receivers may be required for time synchronization.

✔ **Ranging errors:** Although UWB is relatively immune to multipath fading, the key sources of ranging errors in a UWB RTLS are non–line-of-sight propagation and interference from signals from multiple tags and other UWB applications.

Understanding the 802.15.4a standard

Until recently, UWB was mostly a proprietary technology. Recently, however, IEEE approved the 802.15.4a standard for UWB. The IEEE 802.15.4a standard for Wireless Personal Area Networks (WPAN) enables accurate range and distance estimation between communicating devices.

The 802.15.4a standard is crafted to allow the coexistence of coherent and non-coherent UWB radios in the same network. It specifies two optional signaling formats based on *Impulse Radio UWB* and *Chirp Spread Spectrum* (CSS). The Impulse Radio UWB enables ranging capability, whereas the CSS signals can be used only for data communication.

The following list provides the frequency ranges used in impulse radio and CSS:

✔ **Impulse Radio UWB:** 250–750 MHz, 3:244–4:742 GHz, or 5:944–10:234 GHz

✔ **Chirp Spread Spectrum:** 2:4–2:4835 GHz

In IEEE 802.15.4a terminology, the method of ranging is Two-Way Time of Arrival (TW-TOA). This model is very similar to *Round Trip Time* (*RTT*, which I discuss in Chapter 2). In this model, the range between two devices is determined typically by one device sending a message to the other and then the other device turning around and sending the message back.

Using ZigBee

ZigBee is a technological standard created by the ZigBee Alliance for Control and Sensor Networks and is based on IEEE 802.15.4. ZigBee is a simple but flexible protocol that enables high throughput and low latency for low duty-cycle applications. Originally designed to control lights, switches, thermostats, appliances, and the like, ZigBee's self-forming and self-healing mesh network architecture enables its use for an RTLS.

How ZigBee is used for locating

The working principle of a ZigBee RTLS can be described as follows:

1. The first step is to create a ZigBee mesh network. On start-up, the ZigBee coordinator node sets up the PAN network and starts allowing ZigBee routers (that is, location sensors affixed at well-known locations) or ZigBee end devices (tags attached to assets or carried by people) to join to it. Each ZigBee router, on the other hand, tries to join to a ZigBee router that has already joined the PAN or coordinator that is allowing joining. At the end of the mesh formation, each router has a path to the coordinator.

2. The tags (ZigBee end devices) try to join a router or coordinator that is allowing joining.

3. On demand or periodically, the tags take measurements such as TOA and RSSI value of signals from nearby ZigBee routers and send this information to the location engine through the coordinator.

4. The location engine computes the location using trilateration (by converting RSSI or TOA to distance) or fingerprinting (using RSSI). These methods are described in Chapter 2.

The ZigBee system is a self-healing mesh network, which means that it permits data and control messages to be passed from one node to other nodes via multiple paths. This way, ZigBee devices always can speak to the coordinator or location engine even if one path fails.

ZigBee pros and cons

ZigBee enjoys many advantages as an RTLS technology:

✔ **Standards-based:** The ZigBee part of the standard guarantees interoperability of equipment from different manufacturers, which usually results in lower prices.

✔ **Excellent performance in harsh environments:** ZigBee technology relies on IEEE 802.15.4, which has excellent performance in low SNR (signal-to-noise-ratio) environments.

✔ **Fast locating:** ZigBee end devices join or reroute on failures of path very quickly (under 30 milliseconds), enabling fast locating.

✔ **Low cost tags:** ZigBee transceivers are inexpensive.

✔ **Fault tolerant:** ZigBee achieves reliability through mesh networking. If one path stops working, a new path is automatically discovered and used without stopping the system operation.

✔ **High battery life:** ZigBee transceivers are very power efficient as a result of the short working period, low power consumption of communication, and near zero power consumptions in standby mode.

✔ **Large network capacity:** ZigBee supports complex, self-healing mesh networks with as many as 65,000 nodes.

ZigBee has the following disadvantages:

✔ **Limitations of RSSI:** Being in the same spread spectrum as technology like Wi-Fi, ZigBee faces the same limitations with respect to RSSI as Wi-Fi. RSSI is affected by various environmental features, such as obstacles, multipath fading, temperature and humidity variations, opening and closing of doors, furniture relocation, and the presence and mobility of human beings (see the earlier section, "Wi-Fi RTLS pros and cons").

✔ **Interference:** ZigBee operating in ISM bands has same interference issues as Wi-Fi (refer to the "Wi-Fi RTLS pros and cons" section, earlier in this chapter), including potential interference from Wi-Fi networks.

✔ **Line-powered infrastructure:** ZigBee mesh requires that the infrastructure nodes (ZigBee routers) are line-powered because mesh nodes always have to be scanning to pass on the messages from ZigBee end devices or other routers.

✔ **Short range:** ZigBee has a short range that necessitates a high number of ZigBee routers. The penetration of the signal through walls is very limited, so several routers must be placed in every room.

✔ **Security:** Many network attacks discussed in Chapter 15 could be executed against ZigBee RTLS, including jamming, blocking, and denial-of-service attacks.

Because ZigBee operates in the 2.4 GHz unlicensed band, if you have Wi-Fi deployed in your network or other devices operating in 2.4 GHz, you may want to do site surveys to plan the best channels for operation of your ZigBee network.

Using computer vision

Locating systems based on *computer vision* process image data, typically obtained using live cameras, to determine location of assets or people. Image data can take many forms such as still images, video feeds, views from multiple cameras, data from medical scanners, and so on.

ZigBee frequencies

ZigBee operates in three unlicensed frequency bands that include 16 channels at 2.4 GHz, ten channels at 902–928 MHz, and one channel at 868–870 MHz. The maximum data rates for each band are 250 kbps, 40 kbps, and 20 kbps, respectively.

✔ **868.3 MHz:** 1 Channel (Channel 0); operates in Europe, Australia, New Zealand, and the Americas.

✔ **902–928 MHz:** 10 Channels (Channel 1–10 2 MHz gap); operates in Europe, Australia, New Zealand, and the Americas.

✔ **2405–2480 MHz:** 16 Channels (Channel 11–26 5 MHz gap); operates worldwide.

Like Wi-Fi, ZigBee uses Direct Sequence Spread Spectrum (DSSS). O-QPSK modulation with a 32 PN-code length and an RF bandwidth of 2 MHz is used in the 2.4 GHz band. BPSK modulation with a 15 PN-code length (and an RF bandwidth of 600 kHz in Europe and 1200 kHz in North America) is used in the 868.3 and 902–928 MHz bands.

Computer-vision locating is typically used in circumstances where it isn't possible to attach explicit tags to people or assets.

Following are some RTLS applications that use computer vision:

✔ **Drowning-detection systems:** By using computer vision (deploying overhead and/or underwater cameras), you can provide constant pool surveillance, enabling lifeguards to monitor what's happening in the pool and quickly initiate a rescue and save a life. Resuscitation of drowning victims must be initiated as quickly as possible — ideally, within 30 seconds.

✔ **Unusual event detection:** By using computer vision (multiple cameras), you can create a system that detects unusual events and alerts you when the visual patterns are significantly different from the baseline. This feature can be used in security applications in curfew areas, banks, and government buildings, for example.

✔ **Human tracking:** Human tracking can be used to track criminals in crowded environments, as well as to locate lost people, such as children in a shopping mall.

✔ **Vehicle tracking and traffic surveillance:** Video image processing can yield traffic parameters such as flow, velocity, lane changes, and vehicle trajectories, providing better traffic information.

Other applications of computer vision

Image and video content understanding and analysis methods have been studied by many researchers for over two decades, and the recent enormous improvements in visible cameras, infrared cameras, and medical imagers have led to an explosion of applications. One of the most prominent applications of computer vision is in the medical field where image data is in the form of ultrasonic images, microscopy images, X-ray images, colonoscopic video, and so on. Computer vision is used to detect and locate benign and malignant tumors in x-ray images, digital mammograms, colonoscopic videos, and so on.

Some other examples of using computer vision include robotic control of an unmanned lunar rover; locating fruits in trees for robotic harvesting; using surveillance cameras (such as "red-light cameras" to enforce traffic laws); and fire and flame detection in auditoriums, tunnels, and atriums where conventional smoke and fire detectors can't be used.

How computer vision works

The working principle of computer vision locating systems can be described as follows:

1. Image data is sent from cameras (still image cameras or video cameras) to the location engine or application software over the network (wired or wireless).

   No tags are required.

2. The location engine or application software analyzes received images from one or more cameras for specific patterns. The accuracy of the location engine depends upon the image resolution, environmental conditions, and specifications of the applications. Depending upon the application, the accuracy can be of an order of a few centimeters.

The pattern-recognition methods used by the location engine in computer-vision systems have to be designed according to the specific application. The algorithms that work for human-head detection aren't the same as the ones needed for fire detection, for example.

Computer vision pros and cons

Several commercial and non-commercial systems perform computer-vision locating, but the generic vision problem is far from being solved. No existing system can come close to emulating the capabilities of a human, however specialized the tasks that computer vision can accomplish.

Computer vision–based locating systems offer several advantages:

- ✔ **No tags needed:** Because information about the object is extracted from the image itself, no tags need to be attached to assets or carried by people.

- ✔ **Security:** Because it is difficult to manipulate live feeds, it isn't easy to defeat the system without raising alerts.

- ✔ **Long range:** Because *depth of field* (the portion of a scene that appears sharp in the image) can be anywhere from a fraction of a millimeter to virtually infinite, computer-vision locating systems can locate at long ranges.

Some of the disadvantages of computer-vision locating systems are as follows:

- ✔ **Environment dependency:** The quality of images by cameras depends upon environmental conditions such as light levels, fog, gases, rain, and so on.

- ✔ **Application dependency:** Because the algorithms used depend upon the application, design and development efforts are required for every new application. What works in one environment for one application may not work for another.

- ✔ **Camera infrastructure:** An extensive number of cameras might be required to achieve locating. The number of cameras needed depends upon the requirements and specifications of the application.

- ✔ **High network bandwidth:** Transferring high-resolution images or video from cameras to the location engine requires large network bandwidth. This may necessitate adding more back-end network infrastructure capacity.

- ✔ **Expensive to deploy:** The deployment of a computer-vision locating system typically involves high-resolution cameras, installation costs, and high-performance computers for real-time processing of images received.

Using acoustic locating systems

Acoustic locating systems are based on sounds, usually listening for well-defined sounds such as weapon fire. Because sound waves can bend (refract) as they travel through the air, and because sound can travel as far as a mile or more from its origin, acoustic location systems can detect sound events at long ranges and don't require line of sight. Following are some sample applications that use an acoustic RTLS:

✔ **Counter-sniper systems:** Military personnel are often most vulnerable in stationary environments, and locating snipers can prevent casualties.

✔ **Public safety:** Knowing the precise locations of incidents involving guns speeds police response and enables targeted enforcement strategies, from manned surveillance to violence-suppression and threat-reduction missions.

How an acoustic RTLS works

In the acoustic locating systems, the working principle can be described as follows:

1. Typically no tag is used and the object located is the object creating the sound. A tag that acts as an active source of sound can be attached, however.

2. Sensors (either wireless or wired) listen for sounds constantly and report any acoustic anomalies or patterns to the location engine. For gunfire detection, for example, these sensors are deployed across wide areas (say 20–25 per square mile).

3. The location engine performs acoustic trilateration based on the locations of the acoustic sensors and the sound volumes.

The accuracy of locating depends upon the distance between the acoustic sensors and the source of sound, as well as the amount of noise in the environment at the time the sound is generated. The accuracy of locating in gunfire-locating systems is typically of the order of 25 meters.

Acoustic RTLS pros and cons

Although acoustic locating has limited applications, it offers several advantages:

✔ **No tags needed:** Because the object located is the object creating the sound, typically no tags need to be attached to assets or carried by people. This is especially useful for applications such as detecting the location of gunfire.

✔ **No line of sight needed:** Because sound waves can bend (refract) as they travel through the air, sensors can detect the noise around corners and over hills.

✔ **Long range:** Because sound can travel as far as a mile or more from its origin, acoustic location systems can detect sound events at long ranges.

Some of the disadvantages of acoustic locating systems are as follows:

- ✔ **Interference:** Because environments contain a variety of noises, acoustic locating can be used only for special sounds (and acoustic anomalies can give false positives).

- ✔ **Scalability:** Acoustic locating can be used only when there are a very limited number of sound sources.

- ✔ **Easy to defeat:** Acoustic locating for special applications such as gunfire detection may be defeated by use of suppressors or silencers.

- ✔ **Limited applications:** Using tags that make sounds isn't a suitable choice for asset or people locating.

Using building illumination

Because light conditions are dependent on location, researchers have been investigating the use of solar cell–based tags to track light levels and do real-time locating. The key advantage, as in powerline positioning (as discussed in Chapter 9) and dead reckoning (as discussed earlier in this chapter), is that the technology is minimally invasive. The location is computed by using the existing lighting levels in the building.

How building illumination works

The working principle of an illumination-based locating system can be described as follows:

1. Tags are typically dual mode — they have a solar cell and back-end network interface (such as Wi-Fi).

2. The solar cells detect changes in the radiant intensity and communicate this information to the location engine by using the back-end network interface.

3. The patterns of radiant intensity in different parts of the building are recorded as part of the installation, and the database of these patterns is made available to the location engine.

4. The location engine determines the location of tag as the location in the database that has the closest match to the radiant intensity reported by the tag.

Using solar cells as batteries

Because solar cells could be used to not only detect and report the radiant intensity, but also to harvest solar energy, it is possible to have tags that don't need additional batteries. However, without a battery, it may not be possible to use a generic back-end network interface (such as Wi-Fi) because the solar cell lacks enough power.

The tags in that case can do only low-power, small-range RF transmissions. In this scenario, additional RF receivers can be used that could be carried by people, attached to assets, or installed throughout the facility. The location of these RF receivers isn't important, because it just communicates the information reported from the solar tag to the location engine.

Building illumination pros and cons

Extracting context data from indoor lights has several advantages:

- ✔ **Easy installation:** No wiring, cabling, or installation of additional infrastructure (location sensors) is needed.

- ✔ **Small tags:** Solar cells can be produced in thin (less than 1 millimeter thick), low-weight layers.

- ✔ **No batteries needed:** Because the solar cells are energy-harvesting devices, the tags have no battery-life issues.

Because the locating accuracy in this approach depends upon the scene analysis based on the radiant intensity reported by tags, the location engine algorithms face many challenges, including the following:

- ✔ **Natural light:** The presence of natural light from windows may significantly influence radiant intensity reported by the tags.

- ✔ **Uniform intensity:** Many buildings, such as hospitals, have similar lighting conditions in many rooms. The location engine might not be able to differentiate the patterns of radiant intensity from one room to another.

- ✔ **Orientation of tags:** The radiant intensity as seen by tags also depends upon the orientation of tags.

- ✔ **Dynamic environments:** The radiant intensity as seen by tags is also affected by shadows, reflections, and movement in dynamic environments.

- ✔ **Change of lights:** Because the location engine depends upon prerecorded radiant intensities, changes in types and power of lights have to be taken into account.

- ✔ **Aging of lights:** As lights become old or dirt accumulates, the radiant intensity of their light reduces. You may not notice these reductions, but they can adversely impact the accuracy of the location engine.

Chapter 12

Presence-Based Locating

In This Chapter

▶ Understanding how presence-based locating works

▶ Recognizing the value of presence-based locating

Sometimes, you don't need to know the actual location of a person or asset — just whether that person or asset is in the area. Presence-based locating does just that. It just indicates availability (*yes* or *no*).

When you take inventory of your products, for example, you just need to know whether the product is on the premises. Or for example, after the workday is over, you may want to turn on the security alarms or turn off the lights to conserve energy after you confirm by presence-based locating that all employees have left the building.

Presence-based locating is one of the simplest forms of RTLS and you can use many of the technologies that I discuss in Chapters 8 through 11 to do presence-based locating. In this chapter, I explain how to achieve presence-based locating and offer you some examples of real-world applications of the technology.

Detecting Presence

Presence and *location* are two critical qualities associated with any asset or person. Presence indicates availability (yes or no), and location indicates exact position.

Some applications detect presence by using room- or subroom-level locating. Other applications use precision or choke-point locating, which are at the sharp end of location-based technologies that detect presence. Because detecting presence limits the applications you use to ones that don't need precise or symbolic location, you should try to implement it with a minimal investment and/or minimal infrastructure.

The following sections cover fixed and minimal infrastructure solutions for detecting presence.

Powerline positioning

As discussed in Chapter 9, powerline positioning is an inexpensive way of locating (except for the manual labor cost of fingerprinting). In powerline positioning, you don't need to deploy any location sensors except for a few radio frequency (RF) generators that are connected to power outlets in the building. These power generators add RF in the power lines that can be detected by RF detectors (tags).

For accurate locating, you need to do significant fingerprinting for a powerline-based RTLS, but for presence-based locating, the system can be installed trivially with no or minimal fingerprinting. Whenever a power-positioning tag detects the expected RF in the powerlines around it, it communicates that information to the location engine by using Wi-Fi or any other wireless data network, informing the application that a tag is present in the building.

Wi-Fi

If you have Wi-Fi deployed in the building for data network, you can use Wi-Fi tags to enable presence-based locating. You can't use a Wi-Fi data network for precision locating because the margin of error can be anywhere from 10–30 meters, but you don't care about this range if you are just looking for presence.

For more details on a Wi-Fi RTLS, see Chapter 11.

Building illumination

As I explain in Chapter 11, building illumination can be used to locate tags accurately in the building. In building illumination, you don't need to deploy any location sensors, and solar-based tags are located based on scene analysis (fingerprinting of illumination characteristics in the building).

For precision locating, you need to do significant fingerprinting for a building-illumination–based RTLS. For presence-based locating, however, the system can be installed with limited fingerprinting of entrance and exit areas. The tags continually capture the fingerprints of the surrounding light and send the data back to the location engine via Wi-Fi or any other wireless data network, informing the application when a tag is leaving or entering the building.

Mobile locating

Mobile locating is an RTLS solution in which assets or people are located without making use of fixed infrastructure (location sensors). Tags are attached to assets or carried by people and are located by a mobile device. The location is usually in the form of presence, and range depends upon the technology. Mobile locating can be achieved using many of the technologies that I discuss in Chapters 8 through 11. Here are just a few examples:

- ✔ **Passive RFID (Gen 2, UHF):** In this case, a mobile passive RFID reader is used for locating passive RFID tags. The range is typically up to 10 meters (although it can be more or less based on the antenna size and characteristics). A typical example is checking inventory of passive RFID tagged items on shelves in a retail store.

- ✔ **Passive HF:** In this case, a mobile passive HF reader is used for locating passive HF tags. The range is typically up to 3 meters (although it can be more or less based on the antenna size and characteristics). A typical example is using HF tags on vials of drugs to reduce the likelihood of theft and human error, because the drugs are expensive and can cause severe harm if administered incorrectly.

- ✔ **Wi-Fi:** By exploring the ad-hoc capabilities of Wi-Fi, Wi-Fi tags can be located by any Wi-Fi–enabled handheld. A typical application is locating children or other family members by making use of a Wi-Fi–enabled cell phone.

- ✔ **Active UHF:** In this case, a mobile active UHF reader (for example 915 MHz or 433 MHz) is used for locating active UHF tags. The range is typically 100 meters (although it can be more or less based on the antenna size, shape, and frequency used). A typical mobile application is using active UHF tags attached to assets and a handheld using an active UHF reader module for finding those assets.

Using Presence-Based Applications

A system installed to enable accurate locating may lead to increased stress among employees, because they may become wary about taking breaks or talking to co-workers. A presence-based system is a valuable alternative because exact location of employees isn't known. Employees feel secure, and you can still enable safety, efficiency, and security applications.

The following list describes different applications that track the presence of staff:

- ✔ **Building security:** Usually, you have no way to securely lock a building automatically when the last person in the building leaves for the day. However, if all employees carry active tags that can be detected for presence, the security system can be activated automatically.

- ✔ **Saving energy:** Energy-saving operations — such as turning all the lights off or turning heating or air-conditioning to a more appropriate level — can be done when the last person leaves the building or specific area of the building, such as a conference room.

- ✔ **Evacuation management:** Presence-based locating can enable the evacuation management application to determine if all employees and visitors have left the building (if all employees and visitors wear the tags), as well as if they've assembled in the muster areas.

- ✔ **Visitor control:** By providing tags to visitors, you can ensure that all visitors have left the building and prevent security violations.

The following list discusses different applications to track the presence of assets:

- ✔ **Inventory check:** By placing tags on products and items, you can use presence-based locating to recount (check) the inventory at any time.

- ✔ **Asset inventory:** An asset inventory application is used to determine the inventory levels of assets and take actions based on the results. For example, asset inventory in firetrucks helps the response team to ensure that all desired assets are in the truck at the time of dispatch or later.

- ✔ **Room check:** In hospitals, by attaching tags to various assets, the staff can do a quick check before a room is allocated to a patient to determine whether the room has all the desired equipment (assets), as well as to verify if any asset is present in the room that has been recalled.

Chapter 13

Locating by Associating

In This Chapter

▶ Understanding when locating by associating is necessary

▶ Getting to know the underlying technologies

▶ Exploring the practical applications of locating by associating

When you report the location of a person or object by *associating*, you report the location of that object in reference to another object. For example, a child is near her mother, or the doctor is with a patient. A mountain rescue team searching for avalanche victims, for example, can use handheld computers to locate tags worn by victims. Each rescuer's device reports a victim's position relative to itself and thus helps the rescuers find the victims.

This chapter outlines other practical applications of locating by associating and explores the underlying technologies that make the model work.

Knowing When to Use Associations

Absolute locations can always be transformed into relative locations. If you know the absolute location of A as well as the absolute location of B, you probably can infer the relative distance between the two points, so potentially, you can use all the technologies and solutions that provide absolute locations to perform locating by associating (see Figure 13-1).

Figure 13-1:
Locating
tags
directly.

The locator can find the asset in reference
to his or her position or location.

Absolute locations may not always be available, however, for reasons
including the following:

- **Technology deficiency (coverage holes):** In this scenario, the RTLS
  technology is deployed, but the absolute location is unavailable for at
  least one tag, making it impossible to determine the relative location.
  For example, consider the scenario of two neighbors (each carrying a
  GPS tag) talking to each other, where one of them is standing in the lawn
  and other inside the house. Because GPS doesn't work indoors, the only
  location available is the location of the neighbor in the lawn and you
  can't establish that the two neighbors are standing close to each other.
  Another example is a coverage hole in Wi-Fi deployment that causes one
  of the tags to be not locatable.

- **Large margin of error:** The accuracy of absolute position determined by
  RTLS may be so big that even if absolute locations are available, they're
  not useful for the application. If you're using an RTLS that produces
  accuracy within 3–5 meters, for example, the margin of error for distance
  between the two tags located by the same RTLS is double — that is, 6–10
  meters, which may not be good enough for you to infer that the tags are
  close enough to each other.

- **Inability to deploy RTLS technology for absolute locating:** All locating
  systems compute the locations of tags by using location sensors as fixed
  reference points. In some tactical scenarios, however, deploying those
  sensors may not be practical. Consider a battleground scenario: A
  solution that depends on locating by associating may work reliably to
  determine whether a lieutenant is next to a tank.

One way to address these situations is to use tags that can communicate with each other (as shown in Figure 13-2), and another way is to use technologies where location sensors can be small and wearable (almost like tags). In the example in Figure 13-2, a tag reports information about itself and the tags around it to the location engine. Then the location engine uses this information to determine the relative locations of the tags. Also, if the location engine knows the absolute location of a tag, it can extrapolate the absolute locations of remaining tags. In this model, tags are the mobile points of reference for other tags.

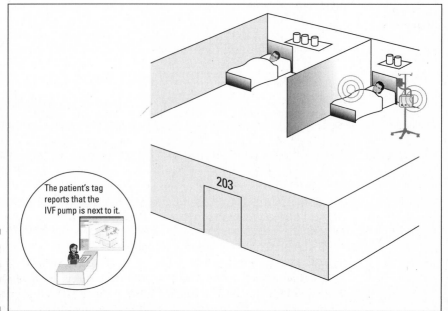

Figure 13-2:
Locating tags indirectly.

Understanding the Underlying Technologies

Typically, to locate by associating, you need tags that can communicate not only information about themselves, but also information about their peer tags. You can use many of the technologies that I discuss in Chapters 8 through 11 to locate by associating. Here are just a few examples:

✔ **Passive RFID:** In this case, association is established between a mobile (typically wearable) RFID reader and passive RFID tags. Assets have passive RFID tags, and people carry mobile passive RFID readers. Because of the short range of the passive RFID reader (up to 10 meters), the location of a passive RFID tag is determined to be the location of the passive RFID reader. The location of the RFID reader is determined manually by the person (carrying the mobile RFID reader) or by embedding another RTLS tag (such as Wi-Fi) in the mobile RFID reader. The Wi-Fi RTLS is used to determine the location of the mobile RFID reader in the latter case. Passive RFID is discussed in detail in Chapter 11.

A typical example for using this model is in warehouses where passive RFID tags are used on palettes and a mobile RFID reader is attached to the forklift truck. As the truck moves around, the locations of all palettes are automatically recorded for inventory and warehouse management.

✔ **Wi-Fi:** In Wi-Fi networks, ad-hoc mode allows Wi-Fi devices to discover and communicate directly with one another in peer-to-peer fashion, without the involvement of central access points. In this mode, a Wi-Fi tag (or device) can see whether another Wi-Fi tag is nearby. To estimate the distance, the tag can make use of RSSI (received signal strength indicator). If needed, the Wi-Fi tags can use Wi-Fi or any other backend network (such as GPRS) to communicate information about these associations (that is, tags in their proximity) to the location engine. A typical example can be doctors and patients all carrying Wi-Fi tags. Not only can the doctors and patients can be individually located using the Wi-Fi RTLS, but an association such as the doctor being next to a patient also can be established. Another example is a mother carrying a Wi-Fi–enabled phone and her child wearing a Wi-Fi tag; the mother can be sure that the child is always within a few meters or present nearby. A third example is a Wi-Fi phone detecting another Wi-Fi phone nearby. See Chapter 11 for more details on how Wi-Fi is used for an RTLS.

✔ **ZigBee:** A key component of the ZigBee protocol is the ability to support mesh networks where each node (ZigBee tag) is self-routing and can connect to other nodes as needed using mesh topology and ad-hoc routing. This association of tags can be used to infer location of any tag with reference to other tags. Using ZigBee for an RTLS is discussed in detail in Chapter 11.

✔ **RuBee:** The design of RuBee technology allows for peer-to-peer communications, not only between tags and routers, but also between tags themselves. In RuBee, the tags could serve as a reader, and this capability can be used to locate tags on the basis of association. For example, it's easy to implement a medical billing application based on how long an IVF pump (with an attached RuBee tag) was located next to a patient (wearing a Rubee tag). Another application is an infant-security application that can ensure that the infant is always next to his or her mother or another authorized person (where all three wear a RuBee tag). RuBee is discussed in detail in Chapter 8.

✔ **Bluetooth:** In Bluetooth, any two Bluetooth-enabled devices (tags) can form an ad-hoc connection and establish a personal area network (known as a *piconet*), which can connect up to eight communicating devices, each identified by its MAC address. And when there is a need for more than eight active devices to form a network, two or more piconets can be connected together into what is known as a *scatternet*. This architecture of Bluetooth enables a location engine to determine which Bluetooth device is in which piconet or scatternet and the associations can be used to locate devices (tags) by association. A typical example is your cell phone's Bluetooth earpiece beeping to inform you when it is moving farther away from the cell phone. Another example is Bluetooth devices installed in aisles of a retail store communicating about the specials and value deals to your Bluetooth-enabled cell phone as you approach or walk through those aisles. See Chapter 11 for more details on how Bluetooth is used for an RTLS.

✔ **Active RFID:** Active RFID transceivers can be used to create proprietary mesh networks where the tags can listen as well as send messages to one another. The associations can also be established between mobile active RFID readers and active RFID tags. For example, typical active RFID tags transmitting at 915 MHz or 433 MHz can use received signal strength from peer tags to establish this association. These active RFID tags can make use of the same frequency (915 MHz or 433 MHz) or another backend network (such as Wi-Fi or GPRS) to communicate information about other tags in their proximity to the location engine. Active RFID is discussed in detail in Chapter 11.

Using Locating by Associating

As knowledge of proximity can be a very effective security indicator, locating by associating is good for applications such as loss or theft prevention. The following sections discuss different applications that use locating by associating.

Security applications

Security and access control are two of the key drivers for an RTLS. And because people and assets are becoming increasingly mobile, there is a strong demand for many security and access control applications using locating by association. The following list describes security applications for locating by associating:

✔ **School trip security system:** When children go for educational and other trips with their school, usually a teacher or guardian is assigned to small

group of students. By having a tag on each child and a tag worn by the guardian of each group, each guardian can ensure that all kids in his or her group are nearby.

✔ **Prisoner transport security:** The transport vehicle can have a tag that is constantly communicating over a wide area network (for example, using GPRS) to the control room and to each prisoner wearing a tag. The vehicle communicates to the control room whether all prisoners are in range and raises an alert if a prisoner moves out of range.

✔ **Authentication:** An asset or person can be allowed into or out of an area without raising an alarm whenever the asset or person is next to an authorized person.

✔ **Tourist group travel management:** For tourist groups, locating by associating can be used to ensure that all members of the group are together. Another example is used by families to ensure all members are accounted for during a visit to a crowded area such as an amusement park.

Safety applications

Locating by associating can be a very powerful tool in safety-critical applications when the risk of personal injury or equipment damage is a primary concern. A large number of accidents in the workplace result from collisions between personnel and machinery. If one tag (on the machinery) can reliably and accurately provide a distance measurement with another tag (on personnel), locating by associating can be used for early warning and prevention of probable collision. This system can help reduce the risk of accidents in the workplace and therefore provide a safer working environment, with lower downtime and liability risks. Applicable scenarios include mining (both aboveground and belowground), loading yards, train terminals, and anywhere else people or light vehicles coexist with heavy mobile machinery.

The following list describes a few safety applications that use locating by associating:

✔ **Mining safety:** A persistent area of concern in mine safety continues to be related to equipment. Many equipment-related injuries and fatalities can be related to haul trucks, belt conveyors, front-end loaders, and other such equipment. Because locating by associating doesn't need any infrastructure, by attaching tags to equipment and by having people carry tags, their proximity can be detected very easily and that can be used to alert workers when they're getting too close to machinery.

✔ **Evacuation management:** In many industrial facilities, it may not be possible or may be too expensive to deploy a locating infrastructure (such as location sensors). In this case, locating by associating can be used to ensure that all people have assembled in muster areas at times of evacuations or drills. A typical way to achieve this is by having a tag in the muster areas that determines the list of people (wearing tags) who have assembled in the muster area and communicating this information to the central server.

✔ **Forklift safety:** Though workplace health and safety requirements have been in place for decades, forklifts frequently cause accidents, some fatal, that result in convictions or fines for companies. Detecting when forklifts and personnel are in close proximity can alert the individuals in time and help prevent injuries. And, because locating by associating doesn't need any infrastructure, this can be achieved anywhere.

Analytic applications

Because locating by associating can provide detailed data on who is near whom, it can be a very powerful tool for analytic applications. In healthcare, for example, it can be used to detect how long a nurse has attended to a patient. It also can be used in restaurants to detect how much time a server spends with a customer. The following list describes some analytic applications for locating by associating:

✔ **Billing:** A patient can be billed accurately based on how much time an IVF pump is on and next to the patient or how much time a caregiver spends with the patient.

✔ **Customer-behavior monitoring:** For example, if tags are attached to shopping carts or carried by shoppers, tags in various aisles can record how much traffic those aisles are seeing. This information can be used by analytic applications to rearrange aisles.

✔ **Museums:** Detecting the amount of time that visitors spend around each piece can help museum staff determine when to change arrangements.

Social-networking applications

Locating by associating can enable a host of social-networking applications. Knowing that a person in your social network is close to you can help you meet new people for dating, business, and friendship. Imagine walking into a meeting, classroom, party, bar, subway station, or airplane and knowing who in your network is already there.

These applications are enabled by you and others in your social network who are carrying tags (typically integrated in Wi-Fi or Bluetooth enabled cell phones). When you move into proximity with someone who is in your social network, you can get alerts (typically on your phone). Because this application can be enabled without needing any external infrastructure, alerts based on your social network can be achieved in any situation.

To address privacy and safety concerns, people can control the visibility of their tags to others, as well as have the choice to opt in or out of the social network.

Part IV
Monitoring Performance and Securing RTLS

The 5th Wave   By Rich Tennant

I located the bear and began testing the vibrating tracking collar over a week ago, but he seems to have left the cave and now I can't find him or the collar anywhere.

In this part . . .

Due to the silent and invisible character of the technology that enables real-time locating of assets or people over the air, an RTLS introduces unique challenges to security and monitoring performance. I wrote this part to add extra emphasis on considering monitoring and security as part of your process when selecting an RTLS.

To dominate the process of achieving whatever things you wish with an RTLS, you need to monitor and measure what really matters. Chapter 14 details the what and the how of monitoring your RTLS, as well as a few metrics that can be used to measure and present performance of your RTLS.

Because good security is as important as good performance, in Chapter 15, I enable you to make the right choice of technology and prepare you should the day arise when you're forced to look for intruders or take countermeasures.

Chapter 14

Measuring RTLS Performance

*W*hen you measure performance, you identify the difference between what you've planned and what actually happens. Monitoring and measuring the performance of an RTLS can be challenging, especially with all the tags, location sensors, technology intricacies, and interdependencies within your organization's infrastructure. There's a lot to keep track of.

This chapter describes the performance aspects you need to monitor and measure in order to determine whether your RTLS is working the way you need it to work.

Defining the Metrics

Having the right performance *metrics* — in this case, a list of performance benchmarks you want to measure — helps you monitor and manage the performance of the RTLS and the application associated with it. Metrics also help you manage how end users are using the application. With the right performance metrics, you can monitor the inner workings of your RTLS closely enough to ensure performance improvements.

Think of the metrics as your scoreboard. For example, if the objective was to reduce losing assets, the number of assets lost per year is a good metric. If the objective was to find assets, average and median time to find an asset are good metrics. And, along with this set of indicators that directly correlate to your application objective, you may want to establish certain other key performance metrics (also known as *Key Performance Indicators, KPI*) to collect all the data you need to take corrective actions, if any are needed.

RTLS metrics are divided into three categories:

- ✔ End-user experience
- ✔ Operational excellence
- ✔ Application objective

And with each metric, you need to define various thresholds:

- ✔ **Target value:** This value represents the value that must be met to say that the objective is met.
- ✔ **Critical value:** This value represents the level that is simply unacceptable.

When you're recording the values of these metrics as you progress, you may also want to measure the worst value, best value, and median value, as well as the number of times those values occur. This scoreboard is your only way to ensure that your RTLS is performing right.

Measuring the End-User Experience

The end-user experience is where, to use an expression, the rubber meets the road. You can be armed with vast amounts of performance metrics on the RTLS infrastructure and application performance, but if you don't know what users are experiencing, you don't have the real performance picture. Although usability of an RTLS application is important, to achieve business value and Return on Investment (ROI), it's critical that you're also monitoring whether users are executing key processes effectively and efficiently.

The following sections cover the key metrics for evaluating an RTLS application's user experience.

Perceived accuracy

Perceived accuracy is the accuracy as observed by a user. Your RTLS may meet the accuracy standards defined in the guidelines from RTLS technology or a vendor that you've selected, but if users feel the system is producing inaccurate results, the system is considered inaccurate.

Users need to believe that accuracy is good or great. If users can identify whether accuracy is great (and if not great, what areas need work), it helps you and your RTLS vendor make the accuracy better.

Consider the example shown in Figure 14-1. Assume that whenever an asset is in the back hallway, the RTLS always returns the asset location, as in room 202. This is within the 5-foot accuracy guideline, but because the only way from the back hall to room 202 is the long route, users will never believe that accuracy is great. This problem needs to be addressed by the RTLS vendor.

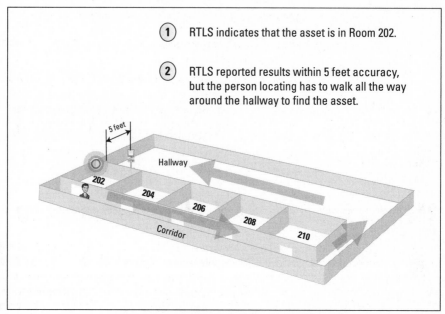

(1) RTLS indicates that the asset is in Room 202.

(2) RTLS reported results within 5 feet accuracy, but the person locating has to walk all the way around the hallway to find the asset.

Figure 14-1: Perceived inaccuracy.

Sometimes, a part of accuracy perception refers to how easily it is to go from a wrong area to a right area. As shown in Figure 14-2, even though the RTLS returns room 202 for all assets in the nearby hallway, users won't mind at all.

Users usually remember the incidents when the RTLS was inaccurate in locating objects or people. This may make users perceive the RTLS as less accurate or unreliable.

If your RTLS provides an obvious way for users to record whether the accuracy was great, you can create a KPI for this as

Percentage accurate = Number of times accurate ÷ Number of times locating attempted

If your RTLS doesn't provide an easy way for users to record accuracy data, however, this indicator should be defined differently. For example, you may define it as *mean time between inaccurate locating*. Users can provide data, such as all instances of inaccurate results, at least once a week.

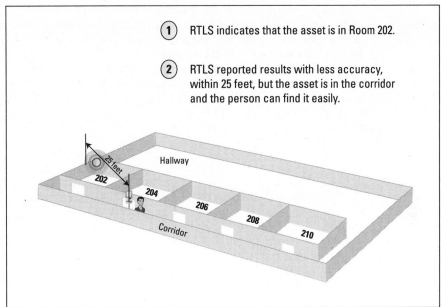

① RTLS indicates that the asset is in Room 202.

② RTLS reported results with less accuracy, within 25 feet, but the asset is in the corridor and the person can find it easily.

Figure 14-2:
Perceived
accuracy.

If you can't measure it, don't define it. Worse than not having a KPI is having a KPI that you can't measure.

Perceived response time

Another key indicator for RTLS performance is the actual response time, as experienced by the end user for all users, all transactions, all locations, all the time. Although this measurement is critical, many organizations fail to consistently capture it.

As with perceived accuracy, if your RTLS provides an obvious way for users to record whether the response time is reasonable, you can create a KPI for this as

> Percentage reasonable response = Number of times reasonable response ÷ Number of times locating attempted

If your RTLS doesn't provide an easy way for users to record response time, this indicator should be defined differently, such as the *rate of irresponsive locating*. Users can provide data, such as when the RTLS returns results slower than is reasonable, at least once a week. You can have a process in which users enter this data every week or so. For example, in an IVF pump locating application in a hospital, the perception of slow and fast has a lot to do with the state of mind at the time and it varies from person to person. You may

think that the response time for locating is good (reasonable) in an area while another user may perceive that it's slow (unreasonable) in the same area. When performance is slow and fast at different times, troubleshoot the performance issue differently than when it's slow all the time. Your Information Technology (IT) department or vendor may spend a great deal of time and energy trying to solve the wrong problem. Metrics that aren't clearly and unambiguously measurable may lead you on a wild goose chase, and you may end up spending too much time, money, and resources troubleshooting the behaviors.

Consistency

Consistency is the capability of an RTLS to produce good and consistent results in ever-changing environments. Consistency is a combination of perceived accuracy and perceived response time. Whether users think results are consistent enough is the most important thing to know.

Tripping and training

Tripping and training are the end-user error metrics. These metrics monitor the critical errors that users encounter. These metrics address the adoption and use of your RTLS application.

The things you want to observe or ask your users are

- ✔ Can users effectively execute the key functions for their specific role or are they struggling to use the application?

- ✔ Have users reached a level of proficiency in which they can execute efficiently or are they struggling to use the application?

- ✔ Where are users having specific issues, and what training is required? Which individual users need specific assistance?

Most of the time, an RTLS application is as good as the efficiency of the users using it. If users need extensive training or have to go through documents or help screens to use the product, they won't be able to use it effectively.

Process intrusion

Process intrusions are the changes in habits and processes that end users have to make when they start using the RTLS. This indicator is crucial. Without it, you may not be able to identify and resolve critical application performance issues that affect the user's ability to perform his or her job function.

The process intrusion metric isn't an obvious (mathematical) indicator. If you communicate with users to establish a rating for this indicator, however, you'll find that most users happily express whether they like the new processes and why.

Gauging Operational Excellency

Most metrics that measure operational excellence for any product are also applicable to an RTLS:

- **Availability:** This metric indicates the historic availability of the RTLS (see the following formula). In a 24/7 operation, the calculation is simply the entire year's potential operating time minus downtime and then divided by the total potential operating time:

   Total seconds in a year – Total down periods in seconds ÷ Total seconds in a year = Percent accuracy

   For example, if you have five minutes of downtime in a year, you have five nines of accuracy:

   $(365 \times 24 \times 60 \times 60) - (5 \times 60) \div (365 \times 24 \times 60 \times 60) = 99.999\%$ accuracy and then $31,536,000 - 300 \div 31,536,000 = .99999$

   Because the RTLS has a large number of components, the *uptime* is basically the sum total time of all elements of the RTLS that have been up and running. In other words, the uptime of each and every tag, location sensor, computer, software process, and so on.

   To be able to determine availability, not only do you need to track any failures when the system became unavailable (or has an extremely slow response) but you also need your RTLS to provide uptime reports (and the time and history of all software or hardware crashes or resets) on all components.

- **Mean time between failures (MTBF):** MTBF is a common metric that establishes the average time between failures. This is a very simple metric and is a good indicator of the product and system quality.

   Like uptime, to be able to determine MTBF, not only do you need to track any failures when the system became unavailable (or had an extremely slow response), but you also need your RTLS to provide you with the uptime and the time and history of all error-level events on all RTLS components.

- **Reliability:** Reliability is a reflection of how dependable the RTLS is. If the failures keep happening (such as the MTBF is low) even though the overall availability of the system may be high, the system may be unacceptable for your application.

✔ **Maintenance cost:** Maintenance cost (or the cost of low quality) is a measure of the number of dollars that are expended rectifying the consequences of bugs. This is typically the sum of labor and material costs. Labor costs cover the work to fix the bug, including the time spent in meetings and calls starting from the initial bug communication to the final bug closure.

Establishing Application Objectives

Key metrics that are application-specific vary along the different applications. The following sections describe the metrics indicators for two sample applications.

Staff emergency response

The purpose of the staff application emergency response is to enable the staff to request assistance whenever they have a personal or medical emergency. The staff carries call-button tags. When they need to summon help, they press a button on the tag.

The metrics specific to this application (and not included as part of operation excellence or end-user experience) are

✔ **The time elapsed since pressing the call button to the time when help comes to the person requesting it.** Median, worst, and best values need to be recorded to figure out what's acceptable and then provide that period as a guideline to the staff.

✔ **Number of times that pressing the call button fails to generate an alert.**

Asset locating

The purpose of the asset-locating application is to enable staff to locate assets (makes sense, right?).

The metrics specific to this application (and not included as part of operation excellence or end-user experience) are

✔ The time elapsed since the staff member actually starts looking for the asset to the time he or she finds the asset.

✔ The number of times an asset couldn't be located by the RTLS but was later found in the facility.

Chapter 15

RTLS Vulnerabilities

An RTLS, by nature of the silent and invisible technology character that enables real-time locating of assets or people over the air, introduces unique challenges to security. What's accessible to an RTLS is potentially accessible to malicious hackers.

Any device implementing the same air interface can potentially access or jam the network. Attacks aren't limited by location or distance. For example, Bluetooth attack tools are known to have several-mile radii, although valid usage scenarios would never attempt such coverage range for Bluetooth.

This chapter provides you with knowledge of the security challenges in an RTLS as well as strategies to counter them.

Evaluating Security Threats

An effective security strategy requires ensuring high availability, integrity trust, and confidentiality. With respect to an RTLS, some of the risks are similar to those of any products, some are exacerbated by wireless connectivity, and some are new.

The heart of the security problem in any RTLS is that the tags aren't tethered with wires. The tags are small and carried by people or attached to assets, and the location sensors or the location engine can locate them with airborne techniques or protocols.

The very ease and flexibility of adding tags that make the RTLS plausible make it highly vulnerable from a security standpoint.

Unauthorized or malicious users may gain access to the tags, access the actual real-time locating system, corrupt the databases, degrade the RTLS performance, or launch attacks that prevent locating from working or providing accurate results.

The following sections outline the risks to security, including attacks on confidentiality, integrity, and network availability. The upcoming section, "Knowing Your Options: Countermeasures," addresses ways to protect your RTLS from attacks.

Availability loss

Loss of RTLS availability involves some form of a denial of service attack, such as jamming the airwaves. The loss of RTLS availability implies that the tags are no longer locatable or that they're not locatable with the expected resolution, accuracy, or performance. Denial of service attacks may be directed at the RTLS in general, at a specific tag, or at a specific location. The following sections cover the four main categories of loss of availability attacks.

Physical security threats

Because an RTLS uses wireless, attackers or intruders don't have to have physical access to attack an RTLS. However, physical access control shouldn't be overlooked. The components of an RTLS, such as tags, are spread all over the facility, and you should have the following means in place to secure those physically:

- **Physical destruction or theft of tags:** Because tags are wireless and small, tags could be destroyed easily. This could be as simple as just removing or discharging the battery or exposing it to physical harms, such as mechanical shocks or chemicals.

- **Physical destruction or theft of location sensors:** Because location sensors are deployed throughout the facility, the sensors could be exposed to many forms of physical destruction. They can be destroyed by changing the orientation or the position of the location sensor, blocking it physically such as putting a metal plate in front of it, or just exposing it to physical harm such as a mechanical shock or chemicals.

- **Tag detachment:** One risk to RTLS security is removing the tag from the asset. At the heart of the RTLS system is the unambiguous identification of the tagged asset by the tag; this type of attack poses a fundamental security problem, even though it may appear trivial at first sight. You may believe that the asset is still present when only the tag is present.

  Another security threat is when the tag from one asset is swapped with that of another asset. You may believe that the asset is still present when it's actually a different asset.

✔ **Physical *power off* of location sensors:** Another risk to an RTLS is the infrastructure of location sensors that could be dependent upon electric power. Environmental events or intruders causing the power off of some location sensors can affect the location accuracy.

Network attacks

Firewalls and intrusion detection systems are standard protection against attacks and unauthorized entries into a company's private network or server. But they aren't enough to counter the security threats that can potentially affect an RTLS:

✔ **Deactivating tags:** These types of attacks render a tag useless without physically destroying it.

Tags (whether active or passive) provide an interface so that users and applications can communicate with the tags wirelessly. By probing and sniffing out communications among tags and other RTLS elements (or by brute force methods, such as cracking passwords, getting access to documentation, and so on), malicious entities may be able to gain control and issue commands to deactivate the tag. Because no physical connection for deactivation is needed, the intruder may actually be quite far from the tag being deactivated. And, depending on the type of deactivation, the location sensors may not be able to identify or even detect the tag's presence.

✔ **Jamming:** Jamming occurs when a malicious hacker deliberately disturbs the air interface between the location sensor and the tag and attacks the communication availability (see Figure 15-1).

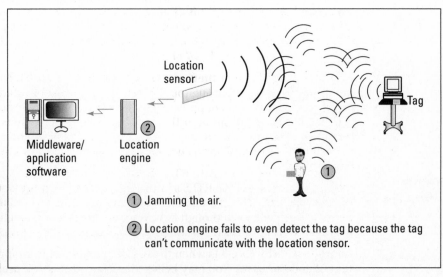

Figure 15-1: Jamming is a network attack.

Location sensor

Middleware/ application software

Location engine

Tag

① Jamming the air.

② Location engine fails to even detect the tag because the tag can't communicate with the location sensor.

The attacker may use physical devices, such as antennas, that can be modulated from a distance to cause radio frequency (RF) jamming. Some jamming tools may use very high RF energy, sufficient to break down the electronics in the tag or the location sensor and make it malfunction completely.

Another way to achieve jamming is through a more passive means, such as *shielding*. For example, when one or more layers of aluminum foil are wrapped around a tag, the tag may be undetectable.

✔ **Blocking:** Blocking is a threat in which malicious hackers create tags that simulate a misbehaving tag (or multiple tags) and confuse the location sensors to fail to respond to the actual tags around this blocker tag. Blocker tags are difficult to detect and can bring down the network.

✔ **Using the man in the middle:** In this attack, the attacker alters a legitimate message between the elements of the RTLS, such as between the tag and the location sensor, the location sensor and the location engine, and so on. The attacker confuses the RTLS by deleting, adding, changing, or reordering the messages (see Figure 15-2).

In one specific attack, the man in the middle can keep the tag from ever talking to the location sensor (or the location engine), resulting in the RTLS system never locating that specific tag.

✔ **Using denial of service:** Besides blocking or jamming, the attacker may prevent or prohibit the normal use of air interface between the tag and the location sensor by *message flooding* (sending repetitive messages) or by *broadcast flooding* (sending messages to all or many destinations at the same time) to the location sensor or the tag. The flooding causes the RTLS to become very slow, to become nonresponsive, or to miss critical packets, which cause it to not locate or to locate inaccurately.

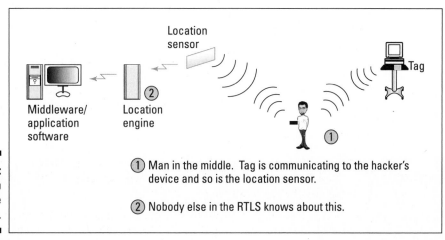

Figure 15-2:
A man in the middle attack.

Location sensor

Tag

Middleware/ application software

Location engine

①

②

① Man in the middle. Tag is communicating to the hacker's device and so is the location sensor.

② Nobody else in the RTLS knows about this.

Many attacks in wired networks, such as *Address Resolution Protocol (ARP) poisoning* (sending an excessively large number of ARP requests to the tag or the sensor, sending fake ARP messages so that an attacker can divert all communication between the tag and the location engine through the attacker's machine, and so on), *Ping of Death* (sending an excessively large number of ping requests to the tag or sensor), and so on, are applicable to the wireless RTLS world, too.

✔ **Masquerading as a location sensor (or location sensor spoofing):** By acting as another location sensor or some other element of the RTLS, the attacker may cause the location engine to not detect or locate a tag.

Product configuration

One of the most common sources of security vulnerabilities is mis-configuration by the users or administrators, exposing the network for attackers or intruders. For example, using factory-default configurations or settings (such as passwords that are often documented and available on vendor Web sites and so on) for some parameters can make it very easy for an intruder to gain access to your network.

Another issue related to product configuration is when the neighboring facilities purchase and use the same RTLS. The two RTLS may interfere or collide with each other, resulting in availability loss (for example, both facilities using the same RF channel for communication between tags and location sensors).

Product quality

The software and hardware quality of the RTLS components is a significant factor. Bugs can expose more vulnerabilities to the attackers.

Integrity loss

Integrity loss involves an active adversary tampering with the data that's being used by the RTLS to compute the tag locations. The loss of RTLS integrity implies that either the tags are being located incorrectly or the information sent to the tags or received from the tags can't be trusted.

Security threats to integrity include

✔ **Masquerading as a tag:** The attacker impersonates a tag, causing the RTLS to get confused and locate the tag incorrectly. This is often referred to as *tag spoofing*.

✔ **Masquerading as a location sensor:** The attacker impersonates a location sensor, causing the RTLS to get confused and locate the tag incorrectly. This is often referred to as *location sensor spoofing*.

✔ **Using the man in the middle attack:** Not just loss of availability, the hackers can use this attack to impact the location accuracy. The tag location accuracy can be selectively affected by altering the messages exchanged between the tag and the location sensors. In this scenario, this attack is called *falsifying the tag data.*

✔ **Distance enlargement and/or reduction:** This type of attack is usually applicable when Received Signal Strength Indicator (RSSI) or Time of Flight (ToF) is used and the attacker impersonates tag signals with different RSSI or ToF measurements.

✔ **Signal synthesis attack:** In this type of attack, a malicious hacker feeds false signals to the receiver (tag or location sensor) to confuse it, thereby, causing the system to fail to locate or to locate incorrectly.

✔ **Replaying:** The attacker monitors messages exchanged and retransmits the messages at a later period, confusing the RTLS.

Confidentiality loss

Confidentiality implies ensuring that information isn't made available or disclosed to unauthorized individuals, entities, or processes.

Confidentiality is, in general, a fundamental security requirement for most organizations; however, due to the shared nature of media used for an RTLS, confidentiality is a more difficult security requirement to meet. Malicious entities may gain unauthorized access to the network through a wireless network made available for the RTLS.

Security threats to confidentiality are

✔ **Eavesdropping:** The attacker monitors the message exchange between the tag and the location sensor or other elements of the RTLS (see Figure 15-3).

The eavesdropped information could, for example, be used to collect private, sensitive information about a person or to collect information from an asset. For example, when an asset tag is connected to an IVF pump and is also used to collect information from that IVF pump, the eavesdropper may get access to information regarding the prescription and dosage.

Another example of this attack is a person listening to the transmissions from the tags or tuning into transmissions between a tag and a location sensor.

✔ **Traffic analysis:** The attacker, in a more subtle way, gains intelligence by monitoring the transmissions from tags for communication patterns. A considerable amount of information is contained in the flow of messages.

For example, the tags may be worn by security guards, and a hacker can determine where the security guards are in the building at any point of time and use it to his advantage. Or in a bank with asset tags for money bags, an attacker can figure out the time periods when more or fewer money bags are in the bank.

✔ **Crackers:** In this attack, the attackers try to *crack* (or figure out) the security keys, if any, exchanged between the tag and the location sensors. By determining the security keys, the attackers get access to the data network (backbone network) of the facility.

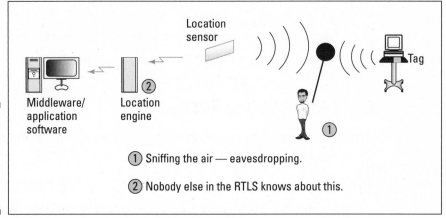

Figure 15-3: Eavesdropping is a threat to confidentiality.

Location sensor

Middleware/ application software

Location engine

Tag

① Sniffing the air — eavesdropping.

② Nobody else in the RTLS knows about this.

Knowing Your Options: Countermeasures

The real impact on RTLS security depends upon the security threats applicable to the technology used and the countermeasures adopted. To physically protect your tags, location sensors, and so on, you must make sure that the proper physical countermeasures (such as barriers, access control systems, and guards) are in place because these are your first line of defense. Other countermeasures include

✔ **Location-based operations:** This is the countermeasure model in which certain operations are allowed from only a specific area. Unless the tag is in that area, the operation is disallowed. Usually, this is easy to achieve when tags have the capability of self-locating.

✔ **Tamper detection:** This is a capability that sends alerts to the RTLS administrator in case the tag or the location sensor is tampered with physically. The idea is to create a tight bond between the tag or the location sensor and the surface it's attached to so that any motion, shock, or removal from the surface can be detected.

Several kinds of tamper detection sensors are available, such as

- *Optical tamper detection:* When the tag is removed from the asset, the sensor in the tag detects light.

- *Bio tamper detection:* The tag monitors a biological component, such as temperature or pulse rate, and when a significant change occurs in any characteristic, alerts are issued. For example, a tag attached to infants for infant-abduction prevention could monitor whether the tag is still on the baby.

- *Magnetic tamper detection:* The tag has a tamper circuit, which gets closed by ferrous contacts that are activated (change positions to close the circuit) in the presence of a magnet. The magnet is included as part of the base of the tag, and opening a tag opens this circuit, signaling a possible attack.

- *Electrical tamper detection:* Some of the adhesive tapes become part of the tag circuit and alert the tag when the tag is separated from the adhesive surface.

- *Strap sensor:* The tag is attached to the asset with a strap, and when the strap is cut, the tag is alerted. The strap has a conductor imbedded within that's in electrical contact with respective terminals on the tag. The tamper-detection circuit on the tag detects when the strap is cut, resulting in an alert.

✔ **Reference tags:** By making use of fixed-reference tags that are always seen by the location sensors, you can assure that all location sensors are active and can read tags.

✔ **Spectrum analyzer:** By installing spectrum analyzers (available in many form factors, such as ceiling sensors), RF signals or noise levels in the environment can be monitored and appropriate alerts can be issued on jamming detection and so on.

✔ **Airwaves noise-level baselines:** The quality of an RTLS depends upon the wireless communication in spite of the noise in the air. By making use of sophisticated software or additional sensors, you can establish levels (or *baselines*) of noise that are typically expected at different times at different locations. And, these software or sensors can send alerts upon significant variations in these baselines.

✔ **Heartbeats:** By making use of periodic checks or *heartbeats* (at whatever periodic levels are acceptable) from all tags and location sensors, the RTLS can provide peace of mind in case a change occurs. These heartbeats are small messages that flow from the tag or the location sensors back to the location engine, indicating that the tag or the location sensor is alive and fine.

✔ **Strong encryption and authentication:** Use the highest possible level of encryption technologies, checksums for all messages exchanged, authentication mechanisms, difficult passwords, and so forth.

Table 15-1 outlines the countermeasures best suited to address each type of threat.

Table 15-1	Threats and Countermeasures
Threat	*Countermeasures*
Physical destruction or theft of tags	Tamper detection, heartbeats
Physical destruction of location sensors	Tamper detection, heartbeats
Detaching tags	Tamper detection
Physical power off of location sensors	Heartbeats, battery-based sensors, or sensors with backup batteries and loss of power as an alert to the administrator or loss of power as a possible input to the RTLS application
Tag deactivation	Strong encryption and authentication methods, allow deactivation only from specific locations
Jamming	Airwaves noise-level baselines, use reference tags
Blocking	Use reference tags
Man in the middle	Airwaves noise-level baselines
Denial of service	Heartbeats, use reference tags
Masquerading as a location sensor	Airwaves noise-level baselines
Masquerading as a tag	Strong encryption and authentication methods
Replay	Strong encryption and authentication methods
Eavesdropping	Strong encryption and authentication methods, avoid transmitting personal data besides IDs
Traffic analysis	No effective countermeasures; strong encryption and authentication methods, avoid transmitting personal data besides IDs
Crackers	Strong encryption and authentication methods

Catching the Attacks

With all the different security threats, the fact that more security vulnerabilities are discovered every day, and the fact that people are actively working on creating attacks, it's pretty tough to keep up with all the latest security countermeasures.

In fact, your RTLS could be under attack and you might not even know it. This isn't a trivial problem. How do you tell whether an attack is going on if there are no obvious signs?

Sometimes your RTLS may have significantly visible symptoms, and sometimes symptoms may be subtle. Sometimes mechanical problems or software bugs can also cause symptoms that may appear like an attack.

To address this issue, you need to have baselines for network performance established. The best way to detect a problem is to keep a constant watch on your baseline and monitor any variation in it. Some RTLS technologies and solutions include the software to characterize this baseline, and that can be an important tool for you to catch the attacks.

Evaluating the Real Security Challenge

Given the large number of security threats and countermeasures and the fact that it's much more difficult to address security after deployment and implementation have occurred, security should be considered from the initial planning stage. You're more likely to make better security decisions at the planning stage.

To help you develop such a plan and understand the seriousness of security risk in your application, security challenges for the following example applications are evaluated in Table 15-2:

✔ **Emergency response:** Consider the environment of a hospital in which push-button tags are provided to staff or patients. They press the push button to summon assistance in the event of an emergency. An attack is high impact if it causes the message from the person requesting emergency assistance to not reach the security guard. An attack is medium impact if the alert is received by a security guard, but instead of one alert, the same alert is received multiple times. It can be annoying, but the purpose of emergency response is still served.

✔ **Animal tracking:** In this application, animals are tagged. The tracking is done to inventory what enters and exits the barn as well as to prevent animals from getting stolen. An attack is high impact if it causes the animal (or inventory) to be stolen.

Table 15-2   Threat Impacts for Sample Applications

Threat	Emergency Response	Animal Tracking
Physical destruction or theft of tags	N/A	High (by animals or a person intending to steal)
Physical destruction of location sensors	High	Low
Detaching tags	N/A	High
Physical power off of location sensors	High	Low
Tag deactivation	High If the attacker can deactivate the tag before the attack, the person can't summon for assistance.	Low
Jamming	High	High
Blocking	High	High
Man in the middle	High	Medium
Denial of service	High	High
Masquerading as a location sensor	High	High
Masquerading as a tag	Medium The attacker can create a large number of false alerts in different parts of the building to divert the security.	High The fact that the animal is stolen will get detected much too late.
Replay	Medium If the alert request by the staff is replayed, it can be annoying; however, the objective of the emergency response application is still met.	N/A

(continued)

Table 15-2 *(continued)*

Threat	Emergency Response	Animal Tracking
Eavesdropping	High If the tags are actively transmitting even when the person isn't summoning for help, the attacker can eavesdrop and locate the person to be attacked.	N/A
Traffic analysis	N/A	N/A
Crackers	N/A	N/A

Part V
The Part of Tens

The 5th Wave By Rich Tennant

"You're in luck — the snake bite kit has an RTLS tag on it."

In this part . . .

Finally, you come to the fun part of this book where the chapters all consist of lists with the ten tips on this, that, and everything in between.

To help you be quick but aware, Chapter 16 lists the ten most common pitfalls that you could run into and what to do about them. Chapter 17 provides you with ten tips for evaluating RTLS vendors. Finally, Chapter 18 provides ten tips about wisely using, recharging, storing, and disposing of the batteries you'll use in tags.

Chapter 16

Ten RTLS Pitfalls

In This Chapter

▶ Overlooking compatibility

▶ Working with an abstract business case

▶ Miscalculating the integration costs

▶ Placing the location sensors ineffectively

▶ Running pilots that omit asset types

▶ Neglecting the business process

▶ Choosing not to deploy enterprise-wide

▶ Underestimating maintenance overhead

▶ Planning for too much too soon

▶ Ignoring privacy concerns

*F*ewer than ten years of rapid and enthusiastic evolution have led to several RTLS technologies based on standards and supported by a large selection of products and services. The promise of an RTLS has become apparent to many inventive solutions providers and users, resulting in companies actively looking to transform their processes with an RTLS. To ensure success, new RTLS users should follow classic guidelines for technology adoption and project management as they move from a pilot to a commercial implementation, and at the same time, use the lessons discovered by early adopters of an RTLS.

In this chapter, I want to make you familiar with the ten most common pitfalls companies and users run into with RTLS implementations. Regardless of your industry and the specific business applications your RTLS will address, by recognizing these pitfalls, you can ensure the short-term and long-term success of your system and truly drive value innovation to achieve both your short-term tactical goals as well as your long-term strategic goals.

Failing to Determine Compatibility

One of the most common pitfalls in RTLS deployment is not spending enough time ensuring compatibility of your choice of technology with the existing technologies in your environment. This is a very crucial issue especially because of the pervasive nature of an RTLS in which tags and location sensors are spread all over the facility.

Look at the compatibility of the RTLS and existing technology with respect to the following:

- ✔ **Networking and other infrastructure technologies:** If a compatibility issue is between the RTLS technology and your network or infrastructure technology, the impact could be network (or infrastructure) disruption, RTLS performance disruption, or both.

  For example, if you're deploying an RTLS solution that makes use of radio frequencies at 2.4 GHz, look for all other systems that use that frequency range, such as Wi-Fi 802.11b. You can prepare a list of all technologies used for the network by talking to your IT (Information Technology) team. For instance, your facility might be using 2.4 GHz video cameras for security, and in that case, it might interfere with the 2.4 GHz you're planning to use for the RTLS.

- ✔ **Technologies used by assets:** Assets today sometimes make use of some technology for data transmission or capture. If a compatibility issue is between the RTLS technology and the technology used by your asset, the impact could be asset functionality degradation, RTLS degradation, or both.

  Prior to attaching tags to assets or installing location sensors, ensure that you're aware of those technologies and there's no such compatibility issue.

  Sometimes one of the reasons isn't interference in technology but just close proximity of two different technologies.

  One approach to address this is by creating the asset list along with the technologies those assets use so that you're aware of this when you select the RTLS technology.

- ✔ **Technologies used by people:** In this era, people use technologies for their personal use and/or for life safety. If a compatibility or interoperability issue is between the RTLS technology and the possible technologies used by people, the impact could be serious.

  Ensure that you've done all due diligence with respect to compatibility and interoperability with devices such as pacemakers and implantable cardiac defibrillators. This is crucial for both giving tags to people and attaching location sensors (because you may have visitors and people with tags in your facility).

One model to address this compatibility issue is by making use of standards-based technologies. Standards-based technologies are usually backed by many groups, technical forums, and organizations and have been examined a lot. However, standardization doesn't guarantee compatibility. On a side note, this isn't the only benefit of standardization: It's inevitably good for the customer because it also drives the creation of a wide selection of products at a lower cost.

Creating an Abstract Business Case

Another common mistake in RTLS deployment is creating an abstract business case that doesn't really reflect operational reality. Don't spend on an RTLS just because your competitors, vendors, or clients are doing it or you think it will help you. Use an RTLS only if you can clearly document the benefits of it for your organization. Different organizations may benefit differently from the same application, and a large gap is usually between the highest theoretical models of what an RTLS can do and the practical realities of the business. For example, if you want to use an RTLS for a staff safety application, clearly document the average, median, and worst case times it took for providing the response before and after using the RTLS.

Measuring performance before and after is essential. Metrics may be hard and harsh, and they may require soul-searching. But if you want to succeed, you have to look at everything. You have to know where you stand before you can make the changes and quantify the benefits.

One way to address this issue is by linking the pilot definition to the business case. This way you can see the business value of an RTLS from the pilot itself and you can make a decision whether the application is a *go* or *no go*.

Underestimating the Integration Cost

A common mistake made by companies is that RTLS integration is perceived as an obstacle to RTLS deployment and is scaled down as much as possible for the initial phase. Although it's a good strategy to plan a smaller RTLS project initially to move forward, by ignoring the larger integration picture, the costs involved in overall integration are underestimated or not planned at all.

For lack of budgets, those integrations either never happen or happen too slowly, resulting in businesses and users living with isolated functionality, multiple instances of the same data, redundant manual activities, higher costs, and inefficient experiences.

I recommend listing all the integration needs, even if you're planning a smaller RTLS project initially. By having all the interface and application needs, you can select the right RTLS solution for you, estimate the potential cost of integration, and establish RTLS deployment phases. If user feedback is positive in initial deployment and there's a measured business benefit, you can demonstrate the success to put forward a strong case for further integration funding. On the other hand, if the feedback is a failure in terms of the expected user benefits compared with the measured results, at least the decision makers can say the project didn't cost very much.

Estimating Location Sensor Deployment Unrealistically

Sometimes, business cases are built on assumptions about how broadly location sensors will be deployed for the desired accuracy. For example, the west wing needs high accuracy, and the remainder of the facility needs low accuracy. These goals and budget plans may not be realistic when the application is deployed because

✔ The accuracy of the RTLS may be less than expected.

✔ You (or the business process) may require higher accuracy than you planned for initially.

One of the reasons that people conclude wrong numbers for location sensors is that they don't test (or *pilot*) in similar environmental conditions to production environment. For example, you may test the location accuracy provided by the RTLS for tags (assets) at orientations and positions that are very well defined, but those may be far different from the tags (assets) at orientations and positions that you might see in real life.

Another reason is that people don't meticulously record data during the pilot phase. This results in perceived good accuracy with fewer location sensors in pilots and then realizing the need for a larger number of location sensors after the full deployment.

One good strategy is to include fixed RTLS tags and automated recording of the tag location as reported by the RTLS. With this automated process, you have objective data of the RTLS accuracy with the given number of location sensors and you can adjust the number of location sensors in the pilot phase. Adding more location sensors or moving location sensors in live deployments is almost a mini-project in itself, and every attempt should be done to avoid that.

Performing Inadequate Testing with Different Types of Assets

Usually due to lack of dedicated resources or just the hassles of procuring all types of assets, pilot tests don't include assets of all types. The impact of this can be

✔ Issues in attaching tags to some assets because of shape, surface, temperature, or environment characteristics.

✔ The RTLS not working on some assets because of interference or some sort of co-existence issues between that asset and the tag technology.

✔ Some assets move at different speeds than others, and the selected technology may not produce the desired results for tags moving at a faster speed.

✔ Some assets vibrate more than others or exhibit other physical characteristics, such as higher temperature, and the selected technology may not produce the desired results for the tags on these assets.

Forgetting about the Business Process

An RTLS is a fascinating technology that provides you with the capability to locate things and people in real time. After you start discovering this and seeing it in action, you may get so engrossed in it that you may almost ignore the business process.

An RTLS isn't about technology but about business-process change. You may have the right technology and the right performance metrics, but by failing to train your users with the new business processes, you may fail to be effective in using the RTLS.

Your users may create new processes to do things — a combination of what they used to do prior to deploying the RTLS, and what they think they should be doing. This just creates overhead and doesn't make your RTLS effective.

To counter this, look for RTLS solutions that require minimum training for the product so that the users can focus on figuring out the new process instead of the new product.

Foregoing Enterprise-Wide Coverage

A system that covers only specific areas or departments impacts your success. User adoption depends on enterprise-wide coverage. Without this, an RTLS will never fully be integrated into the fabric of your organization; also, you can't optimally impact productivity, workflow, or true value innovation.

Although implementing in phases and selecting one area before another is pragmatic, plan for deploying in the whole enterprise. For example, one of the biggest Returns on Investments (ROI) in an RTLS provides reduction in lost, stolen, or otherwise missing equipment. While the assets and people move throughout your entire enterprise, to achieve maximum benefit, your RTLS deployment must cover every square foot. Without coverage that extends everywhere, you simply can't protect or find your assets.

Underestimating Maintenance Efforts

Maintenance considerations, a main factor in both the cost and ongoing success of your deployment, are the ones that are often ignored. Understand the time and money involved in

- **Maintaining your RTLS at peak performance.** It shouldn't take a team of IT professionals to keep the RTLS system running.

- **Replacing batteries in the tags.** This includes not just the cost of batteries but also the time and labor to replace the batteries.

- **Replacing tags or location sensors that aren't functioning properly.**

- **Adding new tags and location sensors.** For this you can leverage your understanding from the pre-pilot/pilot installations.

- **Understanding the needs to tweak location sensors to achieve acceptable results in a live environment.**

- **Monitoring whether all sensors are operational and whether the RTLS system is behaving in general as desired.**

Be sure to understand all maintenance costs upfront and add some maintenance activities as part of your pilot.

Planning for an Overzealous 1.0

Don't plan for the big-bang application as an RTLS 1.0. Select an achievable scope, measure results, and learn from experience — in other words, use the RTLS for basic functionality in the early stages of deployment.

For instance, deploying an RTLS to receive alerts whenever an asset leaves a specific area is a viable 1.0 goal, but at the same time, integrating the RTLS security application with video monitoring in your facility is likely a 2.0 implementation innovation.

Scale up from a foundation of smaller, successful projects. It's easy to get excited and undertake more than can be delivered and monitored.

Failing to Engage Privacy Groups

People adopt what they trust. The RTLS success deeply depends on the people using it. If concerns exist over privacy violations, the RTLS simply will fail.

Engage privacy groups in your area and identify the RTLS technology clearly to your users. Communicate clearly what your intentions are and address any concerns users may have.

Chapter 17

Ten Tips for Selecting an RTLS Vendor

*A*n RTLS isn't a one-time event or a single-use product; it's a way of business. An RTLS isn't something that you install, and some of your employees use it. It changes your processes. An RTLS affects how people do things, how people interact, and how performance is measured in any business division. You'll learn and grow in your RTLS use. You may initially install it for tracking visitors and later use it for employee safety. You may want to use the same system for tracking the dwell times for people and assets to improve efficiency, and you may use the system for tracking assets for asset security.

Because of the potential costs and risks of the wrong decision for something that could be used so widespread and affect your processes, finding the right RTLS vendor is critical. Choosing vendors and replacing them later when they don't work out may mean

✔ Retraining your employees with respect to process changes.

✔ Spending more time in vendor retraining because each vendor must understand your environment.

✔ Adding or removing infrastructure from your walls, ceilings, and environment in general.

✔ Having too many people or vendors working on your systems, which can potentially create more security vulnerabilities, leaks of your business secrets, and so on.

✔ Losing time that can lead your business to not achieve its potential or to lag behind the competition.

This chapter describes ten things that should be part of your RTLS decision matrix for choosing a vendor that meets your needs.

Seeking Critical Mass in an RTLS

An RTLS is a critical piece of infrastructure, just as are door locks, routers, and wired or wireless networks. When you're selecting a vendor, look for one that has *critical mass* in an RTLS, meaning that

✔ **The vendor has a deep enough product portfolio.** Usually, an RTLS deployment consists of tags, location sensors, the location engine, middleware, application software, and the integration with services and systems in your business. A vendor with critical mass has a product portfolio that's deep enough that it can be self-sustaining. For example, if the vendor offers an end-to-end solution (that is, all components of the RTLS, such as tags, location sensors, and so on), you won't have to worry about who will provide the other components.

✔ **The vendor is invested in an RTLS.** Usually, if a vendor has significant revenues from an RTLS and/or is investing significantly in research and development in an RTLS, it indicates that the vendor's invested in an RTLS, and hopefully, you won't have to answer the question, "What will I do if the vendor decides not to stay in the RTLS market?"

Selecting a Strategic Vendor

There's an old adage: "Nobody gets fired for buying IBM." That refers to the glory days when IBM was the top choice in personal computers, and by recommending IBM, you were fairly certain that you wouldn't lose your job because you chose an unknown vendor with an unproven track record. Although times (and IBM) have changed, the concept still holds true: Generally speaking, you're better off selecting a product from a vendor who has a proven track record and a satisfied client base.

However, with an RTLS, no vendor has a long-term proven track record, which makes it more difficult to select a vendor. The RTLS is a fairly new technology, and a large technological evolution has happened in the last seven years. Although some vendors have located at choke points for over ten years, locating precisely indoors is new for almost everyone.

The best strategy is to select a strategic vendor. A *strategic vendor* is interested enough in your organization and the things you're doing to be willing to work with you while you learn and grow. Therefore, you should be a large enough customer to command the vendor's attention.

Finding a strategic vendor is finding a partner with whom you can work to create situations that are beneficial for both organizations. You can honestly discuss your needs, and the vendor can try to find the best solutions for your needs. A strategic vendor is the one who is a long-term fit. The strategic vendor won't just solve your asset-locating needs; it's capable of providing the right solution in case you extend beyond asset locating. The strategic vendor doesn't just provide you a solution, it's your advisor.

Another set of key elements to look for in a strategic partner is vertical specialization and size band coverage. *Vertical specialization,* or *vertical reach,* refers to the extent to which the vendor offers industry-specific solutions and expertise. It's critical to have a vendor who has a strong focus on your vertical market segment. *Size band coverage* refers to the vendor's reach and revenue in your size band. Here are the three company size bands: large enterprises, medium enterprises, and small enterprises. Selecting a vendor who has a focus in your size band will be more motivated and able to provide better support.

Checking End User Sentiment

A natural question when selecting a vendor is "Who uses your product?" What you're probably hoping to hear are the names of companies you've heard of. However, because the RTLS is still a new market, neither the vendor recognition nor the size of the vendor's installed base is high. What this means is that you may not get a lot of references, or even if you get some, many of those are trials and proof of concepts rather than full-fledged deployments. For this reason, it's critical to understand the end user's sentiment of the vendor with respect to the following:

✔ **Product quality:** What's the end user's perception of the product quality? Quality isn't about just having few or no bugs; it's about ease of product use and the vendor's ability to resolve issues as soon as they're discovered.

✔ **Vendor's engagement level:** You're looking for vendors who understand the importance of listening to your end users of an RTLS. For example, if you deployed an RTLS for helping nurses locate equipment, you want a vendor who will have its support and development engineers engaging with nurses to learn and improve. As I mention earlier, an RTLS application isn't a product that's installed, it's about how it's used.

✔ **Capability breadth:** Capability breadth refers to the vendor's ability to go beyond the initial needs and provide consultation, integration, maintenance, and management.

Knowing a Vendor's Technological Focus

Because one RTLS technology doesn't fit all applications, one thing that can be important to evaluate is a vendor's offering of breadth of technologies. *Breadth* refers to the extent to which a vendor meets business requirements across a range of applications and needs. You're not looking for a vendor that supports locating only at room level, locating at choke points, or is hung up on a specific RTLS technology. You want to see openness and vendor capabilities with respect to technology diversity.

One way to look at a vendor's offering of breadth and depth is to find out which technologies and standards the vendor supports.

Try to avoid proprietary technologies! Evaluating a vendor's commitment to common industry standards and support for popular technology platforms is important. Many proprietary technologies have existed in RFID and RTLS space for many years. However, it isn't a hard rule, and sometimes it may not be possible for you to avoid proprietary technologies. Also avoid vendors who don't have an ecosystem of multiple RTLS technologies. If a vendor's offerings include only one specific technology, the vendor may not have you in mind and may try to push that technology for every application need you might have. Using one technology for every RTLS need that you may have might work, but you may not even know the best solution or sacrifices you're making by not using other technologies.

Knowing the Integration and Interoperability Potential

Even though your initial RTLS application may work as a standalone application, it will grow and will need to interact and integrate with many other systems and services in your organization.

You must be aware of the ease and extent with which the RTLS offering can exist, interface, combine, and work with the products, services, and solutions from other vendors. Interoperability and integration is a very important factor because many applications will seek to access, assimilate, analyze, reuse, publish, and act upon real-time information from an RTLS.

Avoid vendors who don't support standard programmatic interfaces.

Asking about Developing Custom Solutions

Let's face it, no two business operations are alike. Businesses differ in the applications, products, protocols, transport layers, operations support systems, physical environment, and processes they use. When you combine any of these aspects, the different combinations any given business might use are unlimited.

Because the choice of an RTLS depends upon your environment, your application, and your application's ability to feed or use real-time information to applications from other vendors, you'll most likely need something customized for a perfect fit into your organization.

Make sure that the company you select has the engineering resources and experience to develop that perfect-fit solution.

Ensuring Smooth Operations with Site Surveys

A *site survey* is the first step in the RTLS deployment by which the surveyor studies the facility to understand your environment. A site survey tells you and the vendor how a specific RTLS technology will perform, what the potential performance issues may be, and where the optimal placement of location sensors is. A site survey also serves as a guide for power considerations, wiring requirements, and installation verification.

Because a site survey is the most important step to ensure the smooth RTLS operation, it's important to know the vendor's capability to perform site surveys.

A few good indicators that a vendor understands the need of site surveys include the vendor using standardized site survey software, its ability to do remote site surveys, and personnel certifications.

Developing a Proof of Concept

A vendor who's motivated to invest in a *proof of concept,* or a trial in your environment, displays the following characteristics:

- ✔ **Confidence in its RTLS offering:** The proof of concept demonstrates in real time the performance, resolution, and accuracy — the three most important claims any RTLS vendor makes.

- ✔ **Motivation to grow with you:** The vendor understands that the time spent during proof of concept helps you and the vendor assess your needs.

Knowing the Shipping Volumes for RTLS Tags

The ability to deliver tag volumes is another critical aspect, especially because most RTLS vendors haven't yet seen high tag volumes and lead times could be significant, often as long as 8–16 weeks.

Be aware of the components and tag manufacturing processes because they can severely limit your ability to get tag volumes in the future.

Asking for Remote Troubleshooting and Updates

What are the vendor's capabilities with respect to remote troubleshooting? This is an important aspect because if the vendor can't remotely troubleshoot, the vendor's response time will be slow. Here are a few quick tests you can do to see if the vendor has remote troubleshooting capabilities:

- ✔ Can a vendor do a remote update (upgrade or downgrade) of any software or firmware?

- ✔ Can a vendor restart any specific software component, all software, or any specific piece of hardware remotely?

✔ Can a vendor remotely check whether any memory leaks have occurred or any processes have died or restarted?

✔ Can a vendor collect all the information its engineers need to debug a particular bug without talking to you over the phone?

✔ Does the vendor have enough logging or tracing capabilities that it can enable? Because excess logging can sometimes change the RTLS response and behavior, it's important that the vendor provides logging capability that can be enabled for any or all components and has multiple levels.

A good indicator that the vendor doesn't have good remote troubleshooting capabilities is when the vendor needs to send support or development engineers to your site for any issue, or the vendor keeps calling you for information.

Chapter 18

Ten Best Ways to Use Batteries

*B*atteries are the lifeblood of an RTLS. Unless you're using passive tags (that have no battery), without batteries, the whole system shuts down.

The key types of batteries that you typically find in RTLS tags (and optionally, location sensors) are: Leclanché, alkaline, nickel-metal hydride (NiMH), lithium, silver oxide, zinc air, and nickel cadmium (NiCd). Some RTLS use *primary* (non-rechargeable) batteries, and others use *secondary* (rechargeable) batteries. This chapter describes ten important points to consider when using batteries for RTLS.

History of batteries

Some historians suggest that the earliest known batteries were invented over 2,000 years ago in Baghdad to electroplate silver on jewelry, and some believe that the Egyptians used batteries to electroplate antimony onto copper over 4,300 years ago. However, the key invention that formed the basis for modern battery technology was done by Alessandro Volta (for whom the *volt* is named) in 1800 in which he demonstrated that an electrical current is generated when metals and chemicals come into contact. And even today, batteries aren't much different. *Batteries* are electrochemical devices that convert chemical energy into electrical energy or vice versa by means of controlled chemical reactions among a set of active chemicals.

Maximizing Battery Life

Batteries lose life because of usage — even when not in use. As the cell ages, the composition and structure of the materials changes, and the result is that the cell capacity deteriorates. You can't do much about usage or aging; however, battery life also depends upon environmental conditions:

✔ **Operating temperature:** Batteries have a limited temperature range over which they work. Attempting to use the battery outside these limits usually results in a permanent degradation in performance or a complete failure.

Keep the operating temperature limits in mind when selecting your RTLS and batteries.

✔ **Operating pressure:** Excessive pressure can cause mechanical failures, such as short circuits between parts, interruptions in the current path, distortion or swelling of the cell case, or in the worst case, actual rupture of the cell casing, within a battery's cells.

Keep the operating pressure limits to which the tags will be subjected to in mind when selecting your RTLS.

Understanding That Not All Batteries Are the Same

In real life, note that all batteries (even from the same manufacturer and the same model) don't deliver their rated capacity. This is due to the manufacturing variations, and these manufacturing variations happen in spite of the fact that manufacturing processes are highly automated and mechanized. The main reason behind this is the spread in the materials' properties used in batteries.

In fact, this manufacturing tolerance spread is one of the key factors behind the wide disparity in the performance of similar batteries from different manufacturers.

For your RTLS application, not only should you use all batteries from the same manufacturer (as well as use the manufacturer and model of batteries recommended by your vendor), but you should also assume extra margins for battery life over and above the recommendations of your vendor.

Disposing of Used Batteries

Unlike single-use batteries, rechargeable batteries continue to make use of potentially toxic heavy metals, such as cadmium, nickel, and lead, which can threaten the environment if not properly discarded. Check with your local government health, solid waste, or recycling department before you consider their disposal.

Rechargeable batteries aren't the only type that can be recycled. Many recycling facilities collect and recycle disposable batteries also.

Under no circumstances should batteries be incinerated because they can explode. If your skin is exposed to an electrolyte, flush with water immediately. If eye exposure occurs, flush with water for 15 minutes and consult a physician immediately.

Knowing the Importance of Cell Casing

The packaging of your battery, or *cell casing,* depends on your RTLS application and more specifically, on the desired size and shape of tags. That is, the size and packaging of the battery dictates the size of tag or vice versa. Here are some common types of cell casings:

- ✔ **Cylindrical cells:** The cylindrical cell is a widely used packaging for batteries because it's easy to manufacture and low in cost. It also provides strong mechanical stability and has good energy density. Some common cylindrical cell batteries are AA, AAA, and so on.

 The drawback of cylindrical cells, however, is that these aren't available in ultra-small size. Because of fixed-cell size, the tag is designed around available cell sizes. If you're working with tags that people wear on their wrists, the cylindrical batteries may not be suitable.

- ✔ **Button (or coin) cells:** The button, or *coin,* cells were developed to reduce size and improve stacking. Built as non-rechargeable or rechargeable, these cells are inexpensive to manufacture and are well-suited for small tag designs. Button or coin cell batteries are commonly used in watches.

✔ **Prismatic cells:** Prismatic cells are contained in a rectangular can — commonly reserved for the lithium battery family. The polymer version is exclusively prismatic. The electrodes are either stacked or in the form of a flattened spiral. They're usually designed to have a very thin profile. Prismatic cell batteries are often used in cell phones.

These batteries are usually custom-made, and no standard cell size exists. Because these cells provide better space utilization, slightly higher manufacturing costs, lower energy density, less mechanical stability, and more vulnerability to swelling aren't considered major disadvantages.

✔ **Pouch cells:** Pouch cells make use of heat-sealable foils, instead of metallic enclosures, and a glass-to-metal electrical-feed trough. This concept allows you to make efficient use of available space and achieve the highest packaging efficiency among all battery packs. Pouch cell batteries are often used when high flexibility in form factor is needed, such as a battery in an ID badge.

Because of the absence of a metal can, the pouch pack is light. Like prismatic cells, these are custom-made, and no standard cell size exists.

Pouch cells typically use lithium polymer chemistries with solid electrolytes, providing a low-cost, flexible construction. The electrodes and the solid electrolyte are usually stacked in layers or laminations and enclosed in a foil envelope.

✔ **Thin film batteries:** Thin film batteries have recently become available in commercial quantities. These batteries usually use lithium polymer chemistries and can be printed on plastics, thin metal foil, or even paper. Thin film batteries are commonly used in semi-passive RFID tags, ID badges, and so on.

Because of their small size, the energy storage and current-carrying capacity of thin film batteries is low; however, because they're bendable, they can be made in any shape or size, operate over a wide temperature range, and have a wide range of uses.

Preventing Sudden or Premature Battery Death

Usually batteries don't exhibit sudden or premature death. The most likely cause of premature battery failure is physical abuse or poor maintenance practices, such as

✔ **Subjecting the battery to excessive vibration or shock.** Not only should you have an understanding of the vibration or the shock the RTLS tag is going to be subjected to, but it would also be of immense value if your tag can somehow report vibration or shock thresholds.

✔ Operating or storing the battery in too high or too low ambient temperatures.

✔ Overcharging or over-discharging as then it may not be possible to recharge it.

✔ Using the wrong charger, such as one designed for charging batteries with different cell chemistries.

Interpreting the Low-Battery Indicator

The purpose of the low-battery indicator is to have an advance indication of the amount of remaining battery life so that you know how much longer you can use the battery before you have to stop and either replace or recharge it.

Battery capacity detectors and indicators are based on one or more battery parameter, such as

✔ **Voltage:** Many batteries exhibit a voltage-drop characteristic in which the voltage drops below a specific threshold, indicating lower remaining battery capacity. The battery voltage is a nonlinear function of everything — temperature, pressure, discharge history, output current, phase of the moon. . . . Although you can detect the low voltage just fine, it's very difficult to predict the remaining capacity of the battery with its terminal voltage alone.

✔ **Current:** The metrics that are used to indicate the maximum capacity of the battery are in the unit of Ah (ampere times hour) or mAh (milli-ampere times hour). Typically, battery manufacturers specify the theoretic total capacity of the battery in mAh. By knowing the current discharge of the battery and the total capacity in mAh, you can compute the theoretical lifetime of the battery using this equation:

$$BLT = C \div I$$

In which *BLT* = battery lifetime, *C* = rated maximum battery capacity in mAh, and *I* = discharge current in milli-amperes.

This battery model allows you to measure the remaining capacity by subtracting the *discharge current* used by the application from the current *capacity*. And, by measuring actual current, you can provide more realistic battery life because that takes into account the deterioration of the useful battery life.

You can actually predict the remaining battery life by monitoring the history of the battery life and employing this history as well as the values of voltage and current as a basis for subsequent predictions of the useful life of the battery.

Most of these indicators predict when the battery is low but may not be able to indicate the actual remaining life because it always depends upon your application.

Whatever approach your RTLS vendor uses to indicate low battery, ensure that there's a sufficient margin (so you have enough time to recharge or change batteries) and your RTLS won't stop suddenly if the battery dies. Another crucial thing is your ability to have access to a centralized dashboard of battery life for all batteries (such as the ones in tags).

Ensuring Battery Safety

As RTLS deployment implies a large number of batteries and battery-based products in your business, be sure users read the safety warnings and follow operating instructions as a handling restriction to ensure safety. Some common safety guidelines for handling batteries are outlined in the following list:

- ✔ **Never disassemble a battery.** The materials inside may be toxic and may damage skin and clothes.

- ✔ **Never attempt to short-circuit a battery.** Doing so can damage the product and generate heat that can cause burns.

- ✔ **Never throw a battery into a fire.** It may explode!

- ✔ **Never throw a battery in water.** It may cause the battery to fail. If there's a potential for exposure to water to your tags, think about enclosures (which I discuss in Chapter 2).

- ✔ **Never insert a battery with the positive and negative poles reversed.** This can cause permanent damage to the battery.

- ✔ **Don't leave the battery in the charger after it's fully charged.**

- ✔ **Avoid mixing old and new batteries together.**

- ✔ **Avoid mixing batteries with differing cell chemistries, such as ordinary dry-cell batteries, NiCd, NiMH batteries, or another manufacturer's batteries.** Differences in various characteristic values and so on can cause damage to the batteries or the product.

- ✔ **Don't put batteries into plastic bags designed to protect components from electrostatic discharge.** These bags are made from conductive material, which could cause the battery to short-circuit.

Make sure you offer periodic training for all RTLS users regarding battery safety.

Preventing Unauthorized Battery Use

For critical applications, authentication may be required to prevent the use of unapproved batteries in the application. This may be to

✔ Avoid allowing a tag to be operated by a non-authorized battery (or by an unauthorized user).

✔ Avoid tag reliability issues that may arise because of battery incompatibility.

✔ Avoid damage to the product's reputation and the brand if inferior or unreliable cells would be used.

Storing Batteries

Batteries start losing capacity right from the moment they leave the factory.

The optimum storage conditions for batteries depend on the active chemicals used in the batteries. During storage the batteries are subject to both self-discharge and possible decomposition of the chemical contents. Over time solvents in the electrolyte may permeate through the seals, causing the electrolyte to dry out and lose its effectiveness. In all cases, these processes are accelerated by heat, and it's wise to store the cells in a cool, benign environment to maximize their shelf life.

The recommended storage temperature for most batteries is 15°C (59°F). However, you should rely on the details from the battery manufacturer about the temperatures and the optimal environment to store your batteries.

Although capacity loss during a battery's storage can't be eliminated, you can minimize the effect by following a few guidelines:

✔ **Store in a cool and dry area.** I recommend refrigeration, but avoid freezers. Also, place batteries in plastic bags for refrigeration to protect against condensation.

✔ **Never fully charge or fully discharge before storage.** Always keep lithium- and nickel-based batteries partially charged because high charge levels hasten the capacity loss and self-discharge on an empty battery may prevent a future recharge.

✔ **Avoid full discharge.** Recharge before the battery is fully discharged because this would wear down the battery unnecessarily.

✔ **Don't stockpile.** Because batteries lose life with time, buy them or start using them as close to the manufacturing date as possible.

 ✔ **Never overcharge.** In nickel batteries, prolonged trickle charge causes crystalline formation, or *memory effect.* Apparently, nickel batteries *remember* the point in their charge cycle when recharging began; during subsequent use they suffer a sudden drop in voltage at that point, as if the battery had been discharged.

Avoiding Bad Charging

Most batteries are damaged by inappropriate battery charging. The life of a recharged battery depends, to a large extent, on the process of charging:

 ✔ **The charger should be appropriate for the chosen battery's cell chemistry because the use of unsuitable chargers can damage the battery.** This is the most important charger requirement.

Don't buy low-quality battery chargers because the charging of batteries is only as good as the charger.

 ✔ **Overcharging severely reduces the battery life.** Your battery specification should indicate the charging time.

 ✔ **The charger design should match the intended-use profile of the battery.** Don't use a charger for one kind of battery chemistry (say, nickel) for a battery of another chemistry (say, lithium).

Don't forget to consider the cost of chargers (and maintenance, such as recharging time, replacement costs of chargers, labor, and storage) with respect to charging as part of your total cost of ownership of your RTLS solution.

One of the important things that you should look for as part of your chargers is the ability to do some kind of message logging (integrating charging time with history) of batteries with your RTLS.

Part VI
Appendixes

The 5th Wave   By Rich Tennant

"I'd like everyone to know that this year's turkey is not only free range and organic, but comes with its own encrypted DNA tracking source code."

In this part . . .

RTLS is a pretty vast topic because it spans across multiple industry segments, multiple technologies, and scores of applications. In this part, I've added these appendixes to provide you some extra material that can be helpful as a reference without distracting you in the main chapters of this book. The appendixes provide a wide range of supplementary topics — essential resources and references, privacy issues, and calculating ROI.

Appendix A

RTLS Resources and References

. . . .

In This Appendix

▶ Finding resources on standardization and technology workforces

▶ Looking up commercial and non-profit resources

▶ Discovering references to RTLS research work

. . . .

*W*hen it comes to an RTLS, a single resource is never enough. This appendix provides a description of essential resources and references for staying up to speed on all things RTLS. I include information on some standardization and dedicated technology workforces, commercial and non-profit organizations dealing with an RTLS, as well as references to many specific RTLS applications and research works.

Knowing where to find these resources will help you be effective in your RTLS choices.

Standardization and Dedicated Technology Workforces

A *standard* is a technical document designed to be used as a rule, guideline, or definition. It's built by consensus and is a repeatable way of doing something. Because RTLS has many parts (such as technologies, tags, enclosures for tags, location sensors, middleware, and application integration interfaces), using standards for RTLS choices can provide assurances of performance, reliability, safety, and interoperability.

The following sections list some of the most recognized and popular standardization bodies and technology workforces doing work related to RTLS.

You may also want to participate in meetings, discussions, and events organized by these groups. The immediate benefits come about through accessing technical resources, communicating and networking with peers in the industry, potentially influencing the standard development, and being recognized for participation.

International Organization for Standardization (ISO)

`www.iso.org`

The International Organization for Standardization (ISO) is a developer and publisher of international standards on a variety of subjects in almost all industry sectors, as well as cross-sectors. ISO is a network of national standards institutes of 160 countries working in partnership with international organizations, governments, industries, and business and consumer representatives.

Refer to `www.iso.org` for more information on standards. The ISO Web site also has information on standards under development or standards that have been withdrawn that can be handy because you may not want to invest in something that is not yet a standard or is no longer a standard.

IEEE

`www.ieee.org`

As stated on its Web site, the IEEE is a non-profit organization that's the world's leading professional association for technology advancement.

IEEE is engaged in development of standards in areas ranging from aerospace systems, computers and telecommunications, to biomedical engineering, electric power and consumer electronics. And the most obvious benefit of accessing IEEE is accessing technical resources related to 802.2, the family of IEEE standards dealing with local area networks and metropolitan area networks. Examples are 802.11 (wireless), 802.15 (wireless personal area networks), and 802.16 (broadband wireless metropolitan area networks).

However, you can also benefit from the knowledge of peers because IEEE is also engaged in enabling the exchange of technical knowledge such as white papers, education activities, technical activities, and so on.

EPCglobal

`www.epcglobalinc.org`

If you want to know anything related to RFID, EPCglobal is your one-stop Web site. EPCglobal is a neutral, consensus-based, not-for-profit standards organization. As stated on its Web site, EPCglobal is leading the development of

industry-driven standards for the Electronic Product Code (EPC) to support the use of radio frequency identification (RFID) in today's fast-moving, information rich, trading networks.

By associating with EPCglobal, you can access information about standards and work in progress for RFID tags and application programming interfaces, as well as link with other companies to create pilots and learn about their experiences.

AIM Global

www.aimglobal.org

As stated on its Web site, AIM Global has actively led the way in industry standards, education, and outreach since 1972, and is the international trade association representing automatic identification and mobility technology solution providers.

Through the AIM Global Web site, you can identify the right applications using Automatic Identification and RFID, influence the direction of markets and standards, as well as connect with new opportunities around the world.

Ubiquitous ID Center

www.uidcenter.org

As stated on its Web site, the Ubiquitous ID Center was set up to establish and popularize the core technology for automatically identifying physical objects and locations and to work toward the ultimate objective of realizing a ubiquitous computing environment.

Because the two key elements of an RTLS are ID and location, by connecting with Ubiquitous ID Center, you can learn as well as be involved in designing Ubiquitous ID technology specifications, participate in the committees and working groups, access specifications before they are released, and more.

Wi-Fi Alliance

www.wi-fi.org

As stated on its Web site, Wi-Fi Alliance was formed in 1999 by several industry leaders who came together to form a global, non-profit organization with the goal of driving the adoption of a single worldwide-accepted standard for high-speed wireless local area networking.

Because Wi-Fi based RTLS is riding the wave of Wi-Fi, through Wi-Fi Alliance, you can stay on top of new developments. Wi-Fi Alliance enables you to learn about the certified Wi-Fi products; get your products certified; find customers; build partnerships; influence working groups on the issues that impact your business; and keep up to date with regulatory updates, market intelligence, road maps, and so on.

ZigBee Alliance

`www.zigbee.org`

As stated on its Web site, the ZigBee Alliance is an association of companies working together to enable reliable, cost-effective, low-power, wirelessly networked, monitoring, and control products based on an open global standard. ZigBee provides companies with a standards-based wireless platform optimized for the unique needs of remote monitoring and control applications. ZigBee Alliance enables you to stay on top of the ZigBee wireless standard, learn how it works and is used for RTLS, and even influence the standards as well as network with vendors and customers.

3GPP

`www.3gpp.org`

The 3rd Generation Partnership Project (3GPP) is a collaboration agreement that was established in December 1998. The collaboration agreement brings together a number of telecommunications standards bodies, which are known as *Organizational Partners*. The current Organizational Partners are ARIB, CCSA, ETSI, ATIS, TTA, and TTC.

A lot of development is in progress with respect to using cellular technology and cell phones as the ID and means for the RTLS. Through the 3GPP Web site, you can gain access to standards, specifications, technology white papers, and latest developments in 3G mobile systems. You also can learn about periodic meetings and explore opportunities to connect with your peers, vendors, and customers.

CEN

www.cen.eu

The European Committee for Standardization or *Comité Européen de Normalisation (CEN),* is a private non-profit organization whose mission is to foster the European economy in global trading, the welfare of European citizens, and the environment by providing an efficient infrastructure to interested parties for the development, maintenance, and distribution of coherent sets of standards and specifications. For example, CEN has many standards for enclosures that you may want to know for business in Europe when using tags, location sensors, and so on.

IEC

www.iec.ch

As stated on its Web site, the International Electrotechnical Commission (IEC) is the world's leading organization that prepares and publishes international standards for all electrical, electronic, and related technologies — collectively known as *electrotechnology.* IEC has many standards for enclosures that are used for tags, location sensors, and so on.

AIDC 100

www.aidc100.org

As stated on its Web site, AIDC 100 is a not-for-profit, self-sustaining, non-political, international organization of automatic identification and data capture (AIDC) professionals and others who have significantly contributed to the growth and advancement of the industry. AIDC 100 is a technical and business resource, with emphasis on education. AIDC 100 is a resource for vendors, users, potential users, and others seeking unbiased information about AIDC technology and markets. AIDC 100 is a great resource as well because of its emphasis on education and social networking.

Bluetooth SIG

```
www.bluetooth.org
```

As stated on its Web site, the Bluetooth Special Interest Group (SIG) is a privately held, not-for-profit trade association founded in September 1998. Its main tasks are to publish Bluetooth specifications, administer the qualification program, protect the Bluetooth trademarks, and evangelize Bluetooth wireless technology. Bluetooth SIG enables you to stay on top of the Bluetooth wireless standard, learn how it works and is used for RTLS, and influence the standards. As well, you can network with vendors and customers.

Commercial and Non-Profit Organization Resources

Many commercial profit and non-profit organizations provide public services in the area of an RTLS. The following sections list the most popular or useful of the various commercial profit and non-profit organizations.

RFID Tribe

```
www.rfidtribe.org
```

As stated on its Web site, RFID Tribe is a global organization with local chapters and is the world's association for radio frequency identification (RFID) professionals. RFID Tribe serves as an engine for ideas, people, and capital. By associating with RFID Tribe, you can connect with others who have similar RFID interests, technical challenges, and business ventures.

RFID Journal

```
www.rfidjournal.org
```

The *RFID Journal* magazine is an independent media company devoted solely to RFID and its many business applications. *RFID Journal* allows you to take advantage of RFID technologies with timely news, strategic analysis, networking opportunities at events, and in-depth education.

RFID Monthly

www.rfid-monthly.com

As stated on its Web site, the primary objective of *RFID Monthly* and its publications is to educate end-users, industry vendors, and investors on key industry developments and trends through analysis and a concise summary of news flow. *RFID Monthly* offers a monthly newsletter to provide you a concise overview and analysis on latest news in RFID software and hardware, as well as industry commentary.

RFID Product News

www.rfidproductnews.com

As stated on its Web site, *RFID Product News* provides a publication dedicated to the RFID industry, its products, and its successes. *RFID Product News* offers a weekly e-newsletter; an RFID vendor, suppliers, and services directory; a comprehensive RFID event calendar; and webinars to help you keep up to date on the most important advances in the technology, hot trends, new products and services, and efficient RFID solutions.

My Blog

www.theRTLSBlog.com

TheRTLSblog is my blog. The blog's mission is to be a comprehensive market intelligence site on RTLS. I include news about various RTLS applications, technologies, and vendors in an easy-to-search way so that you can find and compare information quickly. Through this blog, you can get in touch with me or share resources with other fellow readers.

RTLS Research Work

Over the past few decades, positioning technologies have received increased attention from various authors, scientists, students, universities, businesses, organizations, and governments. Lots of experiments, new product development, and papers have been written in the RTLS field. A common aim for many of these authors and researchers has been improving the overall accuracy of location identification in different environments.

In the following sections, I include a few references of RTLS-related work conducted by some well-cited authors and researchers as well as some key projects related to RTLS. You can check out more information about these resources on my blog (www.theRTLSBlog.com) or by searching the Internet with your favorite search engine.

Active Badge System

The *Active Badge* location system developed at the Olivetti Research Laboratory (now AT&T Cambridge) is an indoor badge-sensing system. This was an early and significant contribution to the field of RTLS. (To find out more, see "The Active badge Location system," *ACM Transactions on Information Systems,* Volume 10(1), by R. Want, A. Hopper, V. Falcao, and J. Gibbons.)

In this system, the locating infrastructure consists of fixed infrared receivers (location sensors) that communicate with the central server (location engine) and the small infrared badges (tags) carried by users. The badges (tags) emit a globally unique identifier every ten seconds or on-demand. This identifier is captured by the infrared receivers in the proximity. The infrared receivers send this information to the central server (location engine) that computes the tag location.

The location of badge is based on which infrared receiver(s) captured the identifier from the tag. For example, if the receiver in Room 20 reports that it received the tag identifier, you know the tag is in Room 20.

Active BAT Location System

Active Bat is an indoor localization system that was developed by AT&T researchers after Active Badge to provide more accurate locating — it reports accuracy in centimeters.

In this system, the locating infrastructure consists of a grid of ultrasonic receivers (location sensors) in ceiling tiles connected by a wired serial network and a radio transmitter (controller). The bats (tags) are attached to equipment or carried by users and have two components: the ultrasonic transmitter and the radio transceiver.

In this model, bats are located one by one. The radio transmitter (controller) sends a synchronized reset signal to all the ultrasonic receivers using the wired serial network, and then sends a request with radio frequency (RF) to one of the bats. Upon receiving the request, the bat emits an ultrasonic message that is received by the ceiling receivers in the area. The ceiling receivers

measure the time from their synchronized reset to the ultrasonic reception and send this Time of Flight (ToF) to the central server (location engine), which computes the tag location. The location is computed using multilateration in which distance between the receiver to the bat is estimated by using the ToF of the ultrasonic signal.

HiBall Tracking System

HiBall Tracker is an optical locating system developed initially for indoors, exhibiting resolution below 1 mm with excellent stability. The Department of Computer Science at the University of North Carolina, Utah, Brown, and Caltech collaborated on the design and production of this project. (For more information, see "SCAAT: Incremental tracking with incomplete information," *Computer Graphics,* Volume 31, Annual Conference Series, by G. Welch and G. Bishop.)

In this system, the locating infrastructure consists of a panel of infrared LEDs embedded in ceiling tiles (location sensors) that take turns flashing. The tag is a miniature cluster of six optical sensors looking out for the flashing LEDs. The tag's optical sensors are continuously looking for LED sightings. The tag communicates the LED sightings to the central server (location engine) that uses knowledge about the geometry of the tag's cameras to compute the location information. The LED flashes very quickly, which allows precise information to be obtained. This technology provides a high update rate and a low latency — solid tracking even at high speeds. And because of the large number of LEDs, it has a built-in redundancy that overcomes most line-of-sight obstructions.

RADAR

Developed by the Microsoft Research group, RADAR is a building-wide tracking system based on the IEEE 802.11 WaveLAN wireless networking technology. (For details, see "RADAR: An In-building RF-based User Location and Tracking System," *INFOCOM 2000.* Nineteenth Annual Joint Conference of the IEEE Computer and Communications Societies. Proceedings. IEEE, by P. Bahland V. Padmanabhan.)

In this system, the locating infrastructure consists of RF transceivers called the base station (location sensors). Generally, laptops or computers with WLAN 802.11 are used as base stations and are positioned to provide overlapping coverage in the area of interest. The tag is a wireless device transmitting signal.

The base stations record the signal strength and signal-to-noise ratio of the signals that the tags transmit and send this information to the location engine. The location engine uses this data to compute the 2D position within the building by combining empirical measurements (also referred to as *scene analysis* or *fingerprinting*) with signal-propagation modeling (to map signal strength for distance) for multilateration.

LANDMARC

LANDMARC (a prototype system developed by researchers at the Department of Computer Science at the Hong Kong University of Science and Technology and the Department of Computer Science and Engineering at Michigan State University) is an indoor locating system. (For more information, see "LANDMARC: Indoor Location Sensing Using Active RFID," *Special Issue: Pervasive Computing and Communications,* by M. Lionel, L. Yunhao, L. Yincho, and P. Abhishek.)

In this system, the locating infrastructure consists of radio frequency identification (RFID) readers at fixed locations and *reference tags.* Each reader has a predetermined power level, thus defining a certain range in which it can detect RFID tags. By placing readers in fixed locations, the whole region can be divided into a number of sub-regions, in which each sub-region can be identified uniquely by the subset of readers that covers that sub-region. The readers continuously scan (read) for the tags. In this process, it is constantly reading the reference tags.

The tag information gathered from RFID readers is sent to the central server (location engine) by a wired or a Wi-Fi network. As dynamic interferences (such as obstructions, human movement, and so on) alter the power level and detection range of RFID readers, the location engine computes location based not only on the tag power level and detection by RFID readers but also on the detected range from the reference tags in real time. The idea behind these reference tags (used as landmarks — hence the name LANDMARC) is that the reference tags are subject to the same effect in the environment as the tags to be located.

Cricket

Cricket is an indoor locating system that makes use of RF and ultrasound, much like the Active Bat system. (For details, see "The Cricket Location-Support System," In Proceeding for the 6th ACM MOBICOM Conference [Boston, MA], by N. Priyantha, A. Chakraborty, and H. Balakrisnan.)

In this system, the locating infrastructure consists of beacons (location sensors) that are small devices attached to fixed locations within the geographic space. Each beacon transmits a data string that uniquely identifies the location of the area where it is fixed and receivers are embedded in the object (tag) being located. The beacons send information about the space over RF together with an ultrasonic pulse. The tag hears the RF signal, uses the first few bits as training information, and then turns on its ultrasonic receiver. It then listens for the ultrasonic pulse, which will usually arrive a short time later.

The tags use the time difference between the receipt of the first bit of RF information and the ultrasonic signal to determine the distance to the beacon. This approach forces the tags to perform all their own triangulation computations. A randomized algorithm allows multiple uncoordinated beacons to coexist in the same space.

NearMe

NearMe is an indoor locating system that uses passive RFID tags. In this system, instead of locating tags using readers, readers are located using tags.

A grid of inexpensive passive RFID tags (location sensors) is spread all over the building, and room identification (such as room number or room name) is stored in each tag. For example, one passive RFID tag is installed on each wall of every room, and the corresponding room numbers are stored in the tags. A mobile RFID reader (tag) is attached to assets or carried by people and it is continuously reading tags around it.

A typical passive RFID reader has a short range (such as 20 feet) and it can read only a few tags around it. For example, when the RFID reader enters Room 235, it will most likely read one or more tags in that room and the data in each tag returns "Room 235." The RFID reader knows its location. The location of asset or person is the room ID (room number or room name) as read by the RFID readers.

GLONASS

GLONASS, a Soviet counterpart to the U.S. Global Positioning System (GPS), first appeared in 1982, four years after the launch of the first Navstar GPS satellite. The Russian Military Space Forces' Global Navigation Satellite System (GLONASS) is designed to provide instantaneous, high-precision location and speed information to users throughout most of the world. Deployed in nearly circular orbits at an altitude of 19,100 kilometers by proton boosters,

each GLONASS satellite emits navigational signals in a 38-degree cone near 1,250 MHz (L2). GLONASS positional accuracies (95-percent confidence) are claimed to be 100 meters on the surface of the Earth, 150 meters in altitude, and 15 centimeters per second in velocity.

As of this writing, the system isn't fully available; however, it's maintained and remains partially operational with 17 operational satellites.

Galileo

Galileo is a global navigation satellite system that at the time of this writing is being built by the European Union (EU) and the European Space Agency (ESA). This project is an alternative and is complementary to the U.S. Global Positioning System (GPS) and the Russian GLONASS.

Galileo is intended to provide more precise measurements (Galileo will be accurate down to the meter range) than are currently available through GPS or GLONASS, including the height (altitude) above sea level and a better positioning service at high latitudes.

Beidou

Beidou, an experimental satellite navigation and positioning system, is China's first space-based regional navigation and positioning network. Beidou is composed of four satellites. The system provides all-weather, 2D-positioning data for military and civilian users. The network covers most areas of the East Asia region and has both navigation and communication functions. And, China has planned to develop a truly global satellite navigation system known as *Compass* or *Beidou-2,* which will consist of 35 satellites.

Unlike the GPS, GLONASS, and Galileo systems, which use medium Earth-orbit satellites, Beidou uses satellites in geostationary orbit. This means that the system doesn't require a large constellation of satellites, but it also limits the coverage to areas on Earth where the satellites are visible.

Appendix B

Privacy Issues

• •

• •

*I*n an era where privacy is valued, tagging people for the purpose of locating them may be met with resistance. And it's not just about tagging people — sometimes even the fact that a person is using a tagged asset has the potential to violate privacy. Although merits of tracking people and assets are driving RTLS demand, it's crucial that you understand the impact on privacy.

Privacy issues may be grouped roughly into two categories:

✔ **Allowing personal privacy:** Different people, cultures, and nations have a wide variety of expectations about how much privacy a person is entitled to or what constitutes an invasion of privacy. In the simplest form, privacy is the right of an individual who is carrying or using an RTLS tag to be left alone. This relates to the RTLS ability to allow an individual to remain unnoticed, unidentified, or untraceable at any time.

✔ **Protecting personal privacy:** This is the ability of an RTLS system to keep the meaning of the information transmitted between the tag and the reader secure from unintended recipients, such as hackers and criminals.

Because any RTLS application's success depends upon the people who carry or use the RTLS tags, without addressing privacy issues carefully, appropriately, and clearly, you can severely limit the RTLS benefits.

Allowing Personal Privacy

Employees and visitors typically have to relinquish some of their privacy while at the workplace or at a place they're visiting, but how much they must do so can be a contentious issue. RTLS applications, such as providing push-buttons tags so that they can summon for help during a crisis or they can be located during a crisis, make this issue more delicate.

One way to deploy an RTLS in a positive manner is to give people control and allow them to choose how the RTLS will be used. For example:

- **Allowing people to make themselves untraceable at any time.**

- **Allowing people to make themselves traceable only when they see a need.** An example is when they're requesting help in a crisis situation.

- **Allowing people to select what information they want to make available when you're trying to locate them.** For example, the tag tells you the person is a teacher but not that it's Ms. Shanaya.

Another way is to clearly communicate the policy regarding how an RTLS application will allow for personal privacy long before the RTLS investment is made. The success of any RTLS application depends upon the people who carry or use the RTLS tags, so you must be clear about the use of RTLS technology involving them.

In your selection of RTLS technology and application, you must consider and provide the desired levels of privacy. For example, for an emergency response application in which you want to deploy push-button tags, you may want to consider how the locating is done. To many privacy advocates, a push-button tag that's located only when the push button is pressed might be a better choice compared with a push-button tag that's always located and pressing the push button merely identifies the emergency event.

Protecting Personal Privacy

If nothing else, the Internet teaches one lesson — there's a lot of interest in everything we do and how we do it. To your left are businesses and governments that want to use this wealth of information, and to your right are criminals and hackers who want to misuse this information.

Privacy is a boundary you need to define and determine. What may be a personal matter to you may not be a personal matter to somebody else. Privacy is all about you, and you have the opportunity to decide what you want to protect and what you don't care about and can afford it if everyone finds out.

Here are the two key types of personal privacy threats:

- ✓ **Individual:** The threats that identify personal information or specific characteristics and behavior of an individual. For example, Jane Doe is in the library surfing the Internet.

- ✓ **Group:** These threats identify specific characteristics and behavior of a group of people. For example, all 21–30-year-old employees are taking longer breaks and spending more time in the cafeteria.

Whether individual or group, the privacy threats originate because of the following:

- ✓ Unauthorized tag readings
- ✓ Using data in ways other than specified originally

Guarding against unauthorized tag readings

After the RTLS tags are attached to assets or carried by people, the most prominent privacy threat is the possibility of an unauthorized tag reading. This unauthorized reading is achieved by reading the tag directly or eavesdropping. The following privacy threats originate because of unauthorized tag readings:

- ✓ Inventorying
- ✓ Traceability
- ✓ Getting personal data
- ✓ Tracking tag locations
- ✓ Establishing social networks

By making use of RTLS technology in which only encrypted data is exchanged among tags, location sensors, and the network, you can ensure that the unauthorized entities can't listen to the data packets exchanged among tags, location sensors, and the network. The other ways to protect from this type of threat are

- ✓ **Use technologies that can't be listened to from a distance.** For example, infrared signals don't penetrate walls and other architectural elements, so a tag's signal is received only within one room.

- ✓ **Use closed loop/proprietary technologies that can't be listened to.**

✔ **Use sensors and security mechanisms to monitor any intruders (or equipment installed by intruders).** This can be part of physical campus security. The processes to actively find invisible intruders with technology are difficult because most intruders are in receive mode or listening mode.

✔ **Use systems to actively thwart the listening.** For example, installing systems (similar to the ones installed inside the building) on building perimeters to confuse the distant people who are trying to listen for data from inside the facility.

Inventorying

Inventorying relates to the ability of an unauthorized entity to read the identifiers sent by RTLS tags without necessarily being concerned as to what the tag is affixed to or who or what is carrying it.

In other words, just by capturing the signals emitted by an RTLS tag, a third party could trace tag absence, presence, or count, thereby violating the organization's privacy.

For example, if a bank starts using RTLS tags on money bags, the intruder, by means of an unauthorized reading, can detect the presence and number of money bags. And, even if the intruder doesn't know what exact things a bank is attaching tags to, the intruder can correlate that information by watching patterns of tag activity.

Traceability

Traceability is similar to inventorying, but the purpose is to locate the tags, such as where the tag is or has been, without necessarily being concerned as to what the tag is affixed to or who or what is carrying it.

For example, consider a scenario in which an intruder detects the tag activity near the facility's perimeter. With visual aid, the intruder can correlate that the tag activity is around the facility's security guard. Now the intruder, by means of an unauthorized reading, can detect the security guard's location or presence.

Getting personal data

A tag may contain personal information. The tag may contain directly identifiable information, such as a name. The tag also may contain data that functions as a *key* (or pointer) to another database in which more personal information is stored.

The privacy threat in the first case is obvious. Whenever the intruder can read the tag data, the intruder can establish a link between the tag ID and the name of the corresponding person.

The privacy threat in the second case — only a key (or pointer) to another database — is indirect. The privacy of the tag carrier is invaded only when the intruder can access the database with the key. This key itself may also be considered as private data because the key alone identifies the person.

Some precautions to protect from this specific attack include

✔ Don't store any personal data in the tag. For example, don't store employees' names in the tags themselves.

✔ Don't keep the association between the tag and a person fixed (keep it changing, or *dynamic*). For example, instead of having a fixed tag assigned to each employee, employees pick up tags from a box of tags at start of a work day and return the tags at the end.

Tracking tag locations

In a case where the tag is carried by a person, an intruder can track the corresponding person's location by way of an unauthorized reading.

In a case where the tag is attached to an asset, by correlating the asset location and knowing who is authorized to move the asset, the intruder can determine the mover's location whenever the intruder detects the asset's movement.

Establishing social networks

By means of data-mining techniques, you can find correlations between separated tags (assets or people). For example, when two tags exhibit similar movement patterns and spend time together in a facility, you may assume that they're somehow related to each other. This information can be enriched by correlating it with other information collected by other sources.

Regulating the use of data

You may install the RTLS application to collect a specific set of data, but later you or one of your colleagues decides to use the data, tags, or location sensors for another purpose. The new applications, although originally unperceived, may have the potential to violate privacy. You must have some processes for self-regulation in place. A few elements of self-regulation include

✔ Giving clear notices as well as choice and opportunity to consent before changes.

✔ No secret (or hidden) tags or location sensors.

✔ Strict policies for unauthorized information capture or similar irresponsible behaviors.

For example, consider the scenario where the attendants in a nursing home facility are tracked to ensure that they're doing facility rounds and to measure how much time they spend in rounds. Later, someone has another idea (especially because attendants now wear tags when they're on duty) to correlate the tracked data on rounds with other data (such as Internet usage, phone usage, and so on) to analyze what the attendants do when they're not doing their rounds. Although the tracking for rounds may have been acceptable to the attendants at first, using this data to track their every move may be seen as a threat to the attendants' privacy.

Another example of this privacy threat is when most of the collected data is profiled with groups, such as age or race. Although this ensures that personal data still remains anonymous, it nevertheless threatens privacy.

Appendix C

Realizing the Benefits of an RTLS

In This Appendix
▶ Assessing the benefits of using an RTLS
▶ Creating a benefits realization report

*B*efore you invest in an RTLS, it's critical to quantify added business value along its three dimensions of how much, how soon, and how certain. Only then can you justify the expense.

The key to assessing how much and how soon your investments will pay off is to set expectations and define measurements at the beginning, and the key to assessing certainty is understanding the possible risks.

This appendix shows you how to create fact-based benefits realization reports so that you can build a compelling business case to justify investments in an RTLS.

Assessing Benefits

Return on Investment (ROI) is used more or less as a generic term to define the business value of an investment. *Classic ROI* is the ratio of the net gain from the investment divided by its total costs. Classic ROI has appeal because of its compelling simplicity; however, different companies and people use a variety of metrics to measure ROI:

▶ **Classic ROI:** To calculate classic ROI, the benefit (or return) of an investment is divided by the cost of the investment; the result is expressed as a percentage or a ratio.

% ROI = [(Gain from investment – cost of investment) ÷ Cost of investment] × 100

Keep in mind that there isn't one "right" calculation for classic ROI — it can be manipulated easily to suit the user's purposes, and the result can be expressed in many different ways. Classic ROI depends on what you include as returns and costs and what your assumptions are.

Say you can sell $30,000 worth of products with Internet advertisements that cost you $500. A marketing person may just present that the ROI for Internet advertisement is 5,900 percent (a gain of $30,000 for the cost of $500), and you may want to include the total value of all other resources (such as material cost, development cost, packaging cost, and so on) that have been employed to make and sell the product. I also remember seeing one ROI calculation based on reducing three contractors for a year with an assumption of $215,000 loaded labor cost per year, when in reality, the company was paying $110,000 to the contractors with little other costs.

When using this metric, make sure you understand all inputs and assumptions that are used. Also, for RTLS solutions, it may not be easy to match specific returns (such as increased profits) with the specific costs that bring them, and this makes classic ROI less trustworthy as a guide for decision support. Classic ROI also becomes less trustworthy as a useful metric when the cost figures include allocated or indirect costs, which are probably not caused directly by the action or the investment.

✔ **Payback period:** This is defined as the time period required to recover the cost of an investment. In other words, it's the point in time at which benefit savings surpass the total project costs. Payback period helps you compare attractiveness of one investment to another.

✔ **NPV:** Net present value (NPV) of an investment is a way to compare the value of money today with the value of money in the future. A future dollar is worth less than a dollar today because of inflation as well as the fact that money today can be invested to grow. To use the future dollar and today's dollar in the same formula, the future dollar is discounted (for instance, multiplied with a number less than one) to represent its value today. This number is the *discount rate*.

NPV = (Future cash inflows × discount rate) – (Today's cost of investment + [Future cash outflows × discount rate])

Computing NPV requires assuming the discount rate. Discuss this with your finance department to determine the discount rate that you can use for ROI calculations. If capital costs your company 12 percent, you aren't likely to invest that capital for a 10-percent return.

The following table gives an example of an NPV calculation for an RTLS investment that a business is exploring to prevent asset loss (estimated to be $36,000 every year). The investment will cost $100,000, and by preventing asset loss, a savings of $36,000 is expected per year.

Year	Discount Rate (10%)	Cash Flow	NPV of Cash Flow
0	0.909	–$100,000	–$100,000
1	0.826	$36,000	$32,724
2	0.751	$36,000	$29,736
3	0.683	$36,000	$27,036
4	0.621	$36,000	$24,588

The total cash flow is $44,000, and the total NPV of cash flow is $14,084.

This is the most favored formula among finance people because NPV is valuable for ranking one project against another. As you can see without the payback period, NPV isn't informative enough. NPV is usually always used along with the payback period.

✔ **Internal Rate of Return (IRR):** Because business cases must have positive NPV for a go or no-go decision, this is what you expect to *earn* by investing in the project to drive the NPV to zero. In other terms, this is the value another investment would need to generate in order to be equivalent to the cash flows of the investment being considered.

✔ **Total cost of ownership (TCO):** Total cost of ownership is the total up-front and ongoing costs for implementing the RTLS solution. TCO includes

- *Capital expenses:* The investment in systems, software, tags, location sensors, networks, supplies, and so on as needed to deploy and maintain the solution.

- *Implementation costs:* The staff and contract labor necessary to research, purchase, plan, test, and deploy the proposed solution.

- *Ongoing maintenance and support:* The staff and contract labor to manage and support the solution after it's deployed, the additional phone and IT costs, and so on.

- *Operations and contracts:* The recurring fees, leases, time spent with attorneys, storage, electricity costs, and general ongoing maintenance and support costs.

- *Change management and training costs:* The management during a time of change in which you spend time and energy to inform involved parties as well as user-training fees and labor and time spent for the trainings.

TCO should never be used in isolation because it can drive you to seek the lowest cost solution of the project to cover the costs.

All the metrics mentioned in the preceding list are nearly meaningless if you can't predict the real benefit. Benefits of an RTLS come in four forms, which I discuss in the following sections:

- ✔ Direct (hard) benefits
- ✔ Indirect (hard) benefits
- ✔ Soft benefits
- ✔ Quantifiable soft benefits

Direct (hard) RTLS benefits

Direct benefits are the benefits directly tied to the impact of implementing the proposed RTLS solution — a first order, cause and effect. Direct benefits are also known as hard benefits, or tangibles, because these are clearly quantifiable, verifiable, and expressed in monetary units. The following sections describe the different types of direct RTLS benefits.

RTLS business cases fail not because of the lack of direct benefits availability but because of the lack of discovery that those benefits actually exist.

Savings

Savings include removing existing costs as well as (future) cost avoidance opportunities. For example, labor savings, such as the reduced headcount, reduced overtime, and so on; and expense reductions, such as reduced inventory, reduced equipment purchases, and so on.

The following list describes some examples of how RTLS can save different organizations money:

- ✔ **Healthcare:** By deploying the RTLS application to locate IVF pumps, you can reduce the number of IVF pumps leased, rented, or purchased every year because the locating helps

  - Prevent theft and accidental loss.

  - Improve asset utilization because most pumps are used more often and can be located easily and redistributed if more pumps are present in an area than can be used.

- ✔ **Mining:** By deploying the RTLS application to track equipment, you can reduce the time spent in finding shared equipment during shift changes.

- ✔ **Manufacturing:** By deploying the RTLS application to track tools, you can reduce the number of tools and test equipment needed because the locating helps reduce search time.

Revenue benefits

Revenue benefits are the increases in revenue and profit, such as incremental sales, reduced sales cycles, and so on. The following list provides some examples of revenue benefits of using RTLS:

✔ **Healthcare:** By deploying the RTLS application to locate IVF pumps, you can potentially increase billing (charge capture) because the locating helps track precisely when and where the pump is set up and then bill accordingly.

✔ **Healthcare:** By deploying the RTLS application for bed management, you can potentially increase net revenues because you get people in and out of the hospital faster. You can track real-time notifications of patient or bed status (such as occupied, available, assigned, discharge ordered, cleaning, no staff, or not in service, and so on), thereby enabling faster transport of patients, faster housekeeping, and so on.

✔ **Manufacturing:** By deploying the RTLS application to track tools, you can speed up production because locating helps you

- Increase tool utilization
- Decrease production line down-time because surveys are completed automatically

Improvements

The improvements benefits, such as working capital, happen as a result of RTLS solution deployment. The following list provides a couple of examples of the improvements gained by using an RTLS:

✔ **Healthcare:** By deploying the RTLS application to locate IVF pumps, you can improve working capital. For example, fewer hard dollars need to be allocated for new IVF pump purchases every year.

✔ **Manufacturing:** By deploying the RTLS application to track tools, you can reduce turnaround for maintenance operations, allowing for reduced physical working space needs.

Indirect RTLS benefits

Indirect benefits, as the name suggests, are likely to happen as a result of implementation but aren't tied directly to the proposed RTLS implementation. Indirect benefits are hard benefits, or tangibles, because they are clearly quantifiable and expressed in monetary units. Indirect benefits are tangible because they're visible, easily understandable, and consistent with the observer's prior experience. In ROI reports, indirect benefits should be discounted to 30–40 percent because tracking these may be a little more difficult than tracking direct benefits.

Indirect benefits are hard and are subject to opinion. For instance, some people may see the indirect connection and others may not.

The following list provides some examples of the indirect benefits of working with an RTLS in various industries:

- **Healthcare:** By deploying the RTLS application to locate IVF pumps, you can reduce storage and maintenance needs. Less overage to purchase, rent, or lease also results in less space allocated for storage as well as less maintenance needs.

- **Manufacturing:** By deploying the RTLS application to track tools, you can reduce staff's overtime hours by increasing worker productivity. More time can be focused on completing manufacturing tasks rather than searching for parts required for the production process.

Soft RTLS benefits

Soft benefits are the payoffs that are neither quantifiable nor expressible in monetary units. An example is reducing a worker's frustration associated with locating an asset. For example, in a manufacturing facility, a lot of equipment is shared. Imagine every morning a worker has to look for the equipment he or she needs. You can quantify the time spent looking, but you can't quantify the benefit of the worker being able to use the RTLS to immediately locate the tools he or she needs to start working.

The following list describes examples of soft benefits of an RTLS for a few example industries:

- **Healthcare:** By deploying the RTLS application to locate IVF pumps, your staff has to spend less time locating equipment, which increases staff productivity and empowers them to do what they were hired to do.

  By deploying the RTLS application to improve patient flows, you can potentially improve perception and reputation by

  - Reducing patient waiting time

  - Tracking patient progress

- **Mining:** By deploying the RTLS application for worker safety, you can track personnel in hazardous environments in mines, which is crucial in the event of an accident. The company instantly knows which workers are trapped and whether an employee remains motionless in a hazardous area.

✔ **Manufacturing:** By deploying the RTLS application to track tools, you can reduce staff frustration by improving their ability to find shared tools and equipment and to focus on and finish their work.

✔ **Manufacturing:** By deploying the RTLS application to track tools, you can improve customer service and customer appreciation because locating enables

- Direct view of the shop floor, which ends the need to manually retrieve overall status at any stage

- Customers to appreciate the accuracy and timeliness of answers to their inquiries

Satisfaction or happiness in employees ultimately reflects in your product quality, increased market share, and even ultimately, industry-position rankings. Even if these benefits are difficult, or impossible, to accurately predict and link them directly to the bottom line, they may provide even greater value to the corporation.

Quantifiable RTLS soft benefits

After a benefit is labeled *soft,* it's usually a bad idea to include it in your ROI reports. Therefore to justify adding soft benefits, the emphasis should be on *quantified soft benefits.* These are the soft benefits that can be calculated monetarily (with varied degrees of opinion) and included in the business case, but still kept out of the ROI equation.

The things that don't seem measurable actually are. For example, you can claim that improving nurse safety will save $100,000 in replacement costs. Replacement costs include cost of hiring (such as advertising), new employee training, overtime costs (as existing nurses have to work more with one nurse less), and so on.

✔ **Healthcare:** By deploying the RTLS application to locate IVF pumps, you can track whether your staff is ensuring compliance with regulations, such as proper disinfection between uses by different patients. To quantify the benefits of deploying such an application, you can look at your past costs and industry average costs spent in case a violation citation is received.

✔ **Healthcare:** By deploying the RTLS application to monitor patient flows, you can potentially reduce the time your staff has to spend to check the status of rooms and beds. The time the staff spends making phone calls and visits can be quantified and translated into a hard-dollar value.

✔ **Mining:** By deploying the RTLS application for worker safety, you can track personnel in hazardous environments. The value of tracking safety can be quantified by using past and industry average compensations as well as expenses in lawsuits resulting because of the company's inability to locate or help employees who are hurt or killed on the job.

✔ **Manufacturing:** By deploying the RTLS application for tracking tools, you can gather information on the time and number of occasions that people actually work on work orders and look for shared equipment and tools in manual surveys. This time can be quantified and translated into a hard-dollar value.

Another way to convert a soft benefit into a quantified soft benefit is to gather data by asking questions that have a scaled response (such as 1 for not at all likely to 5 for very likely), and then using this data to assign a hard-dollar value to at least one aspect of the soft benefit.

Creating Benefits Realization Reports

After you assess the benefits, you need to present the data to all stakeholders. My advice is to *present the right thing.* This data should provide an accurate representation of the business value of RTLS, appeal to the decision-makers, be believable, and capture the total costs and total investments. Everyone should be able to understand it and also trust it.

Graphs and charts do a better job of communicating than tables with numbers. A benefit realization report for RTLS must include charts and graphs demonstrating the benefits as well as the cash flows from various benefit sources.

Tips for creating reports

Here are some tips on how to create benefit realization reports from the perspectives of those who have done it:

✔ **Solicit input from executives.** Make sure you have input from your executives to uncover the primary concerns of key decision makers. There is no benefit in developing the ROI business case if you don't have general buy-in from key parties involved in the decision-making process.

✔ **Prepare a case study.** Talk to users and people at all levels to understand how productivity can be improved. Go to where the action is. Watch them in action and note it. Individuals at higher levels in an organization may not understand how productivity can be improved at different levels.

✔ **Consider all the options.** It's important that you consider all the alternatives. Don't restrict your focus to one technology or approach.

✔ **Include more than one application.** Include ROI and business case for at least one more application that's deployed after the initial application. The value of incremental additions not only demonstrates that the RTLS can help in many ways, but also the incremental value will make your business case solid.

✔ **Analyze expenses.** Spend time looking for expenses and look for repeatable expenses.

✔ **Be precise.** Use spreadsheets or some other calculation tool. No one appreciates mathematical errors in your report.

✔ **Avoid techno-speak.** It's important that the business case aligns the business requirements with the system's value and that the results are understandable and not incomprehensible techno-speak. With newer technologies like the RTLS, it's pretty easy to get excited about technology and easy to focus on technology in an ROI discussion, such as *This technology is so hot* rather than the real benefits.

✔ **Make sure your ROI benefits are auditable and staggered.** Frequent auditing will help you ensure you're on the right path and help you design the right milestones for your deployment. And, if none of your proposed ROI benefits will hit until the third year of the deployment, it's probably not worth doing. Breaking even can take periods longer than six months, but benefits need to start kicking in sooner.

✔ **Include details.** Include details on all benefit statements, underlying assumptions, parameters, costs, and your ROI methodology. You want to address the most penetrating questions and concerns of those looking to pick apart your report.

✔ **Discount soft benefits.** Soft benefits matter, but underplay them. For instance, don't include them in ROI calculations and don't spend too much time talking about them.

✔ **Include confidence level, net present value, and payback period.** Present your results clearly and concisely, such as *We are 90 percent confident that we can achieve a net present value of $1 million with a payback period of 18 months.*

✔ **Be complete.** Include addendums with press releases and Internet search results for competition or similar businesses that have deployed RTLS solutions for similar or the same problems.

Capturing risks to ROI

Like any other initiative, RTLS solution proposal initiatives have different risks associated with them, and this should be called out as part of the analysis.

Having the risk analysis helps you make your analysis realistic, and your chances to realize the full benefits you're expecting increase. Here are a few things you should consider during your risk analysis:

- **Underestimating training costs:** Nothing ruins ROI faster than lack of users using it. One way around this is to ensure easy or obvious use of an RTLS application. If people don't need much training to use it, they probably will use it. See if you can fit the training card of your RTLS application on a credit card–size document that can fit in wallets.

- **Overvaluing the soft benefits:** Don't become emotionally attached to the soft benefits. Soft stuff is soft stuff. Things change, environments change, applications change, and perspectives change. Keep the soft benefits out of your ROI and interpersonal discussions.

- **Leasing models:** Leasing models confuse the ROI calculations because costs don't appear much, and you may select a wrong RTLS technology or solution that may seem to be more affordable. Try not to compare a leasing model against a capital model; rather, translate one of those to the other.

- **Using faulty reasoning:** Verify all your assumptions and benefits before you include them in your ROI report. You should thoroughly test your analysis and its underlying assumptions. In fact, you should have at least one user in mind who would actually receive the benefit that you're describing in your analysis.

- **Overlooking decision-making participants:** Because each stakeholder brings a unique perspective to the decision making, it's crucial for you to know and involve all the decision-making participants. Not only does their involvement help you to build a solid business case, but it may actually prevent you from overlooking some benefits that can come to the table.

Index